End-to-End Automation with Kubernetes and Crossplane

Develop a control plane-based platform for unified infrastructure, services, and application automation

Arun Ramakani

BIRMINGHAM—MUMBAI

End-to-End Automation with Kubernetes and Crossplane

Copyright © 2022 Packt Publishing

Group Product Manager: Rahul Nair
Publishing Product Manager: Niranjan Naikwadi
Senior Editor: Athikho Sapuni Rishana
Technical Editor: Rajat Sharma
Copy Editor: Safis Editing
Project Coordinator: Ajesh Devavaram
Proofreader: Safis Editing
Indexer: Rekha Nair
Production Designer: Shyam Sundar Korumilli
Marketing Coordinator: Nimisha Dua
Senior Marketing Coordinator: Sanjana Gupta

First published: July 2022

Production reference: 1060722

Published by Packt Publishing Ltd.
Livery Place
35 Livery Street
Birmingham
B3 2PB, UK.

ISBN 978-1-80181-154-5

www.packt.com

To my parents, for their commitment to guiding me on the right career path at every little step. To my wife and son, for being very supportive and sacrificing their needs to assist me on this book journey. Finally, to my close friend and buddy, Prabu Soundarrajan, who helped shape my technical perception for this book and beyond.

– Arun Ramakani

Contributors

About the author

Arun Ramakani is a passionate distributed platform development and technology blogging expert living in Dubai, a dynamic city where many cultures meet. He is currently working as a technology architect at PwC, specializing in evolutionary architecture practices, Kubernetes DevOps, cloud-native apps, and microservices. He has over a decade of experience in working with a variety of different technologies, domains, and teams. He has been part of many digital transformation journeys in the last few years. This book is inspired by one of his recent works. He is enthusiastic about learning in public and committed to helping individuals in their cloud-native learning journeys.

I want to thank the people who have been close to me and supported me, especially my wife, son, and parents. I would like to thank Kelsey Hightower, who inspired me to do some significant work around the Kubernetes ecosystem and mentoring me during the office hours. Finally, I want to thank everyone in the Crossplane community for providing me with inspiration for lots of the ideas covered in the book.

About the reviewer

Werner Dijkerman is a freelancing cloud, (certified) Kubernetes, and DevOps engineer, currently focused and working on cloud-native solutions and tools such as AWS, Ansible, Kubernetes, and Terraform. His focus is on infrastructure as code and monitoring the correct "thing" with tools such as Zabbix, Prometheus, and the ELK stack. Big thanks, hugs, and shoutouts to Anca Borodi, Theo Punter, and the rest of the team at COERA!

Table of Contents

Part 2: Building a Modern Infrastructure Platform

3

Automating Infrastructure with Crossplane

4

Composing Infrastructure with Crossplane

5

Exploring Infrastructure Platform Patterns

6

More Crossplane Patterns

7

Extending and Scaling Crossplane

Part 3: Configuration Management Tools and Recipes

8

Knowing the Trade-offs

9

Using Helm, Kustomize, and KubeVela

10
Onboarding Applications with Crossplane

11
Driving the Platform Adoption

Index

Other Books You May Enjoy

Preface

In the last few years, countless organizations have taken advantage of the disruptive application deployment operating model provided by Kubernetes. With the launch of Crossplane, the same benefits are coming to the world of infrastructure provisioning and management. The limitations of infrastructure as code, with respect to drift management, role-based access control, team collaboration, and weak contracts, have made people move toward control plane-based infrastructure automation, but setting it up requires a lot of know-how and effort.

This book will cover a detailed journey to building a control plane-based infrastructure automation platform with Kubernetes and Crossplane. The cloud-native landscape has an overwhelming list of configuration management tools, which can make it difficult to analyze and choose the right one. This book will guide cloud-native practitioners to select the right tools for Kubernetes configuration management that best suit the use case. You'll learn about configuration management with hands-on modules built on popular configuration management tools, such as Helm, Kustomize, Argo, and KubeVela. The hands-on examples will be guides that you can directly use in your day-to-day work.

By the end of this DevOps book, you'll have be well versed in building a modern infrastructure automation platform to unify application and infrastructure automation.

Who this book is for

This book is for cloud architects, platform engineers, infrastructure or application operators, and Kubernetes enthusiasts interested in simplifying infrastructure and application automation. A basic understanding of Kubernetes and its building blocks, such as Pod, Deployment, Service, and namespace, is needed before you can get started with this book.

What this book covers

Chapter 1, Introducing the New Operating Model, discusses how, for many people, Kubernetes is all about container orchestration. But Kubernetes is much more than that. Understanding the deciding factors on why Kubernetes disrupted the day 1 and day 2 IT operations is key to successful adoption and optimum usage.

Chapter 2, Examining the State of Infrastructure Automation, exposes the limitations of infrastructure as code and proposes control plane-based infrastructure automation as the new-age automation concept using Crossplane and Kubernetes.

Chapter 3, Automating Infrastructure with Crossplane, helps us to understand how to set up a Crossplane cluster, discusses its architecture, and explains how to use it as a vanilla flavor for infrastructure automation.

Chapter 4, Composing Infrastructure with Crossplane, helps us to understand composing, a powerful construct of Crossplane that can help us to create new infrastructure abstractions. These abstractions can be our custom Kubernetes-based cloud APIs with the organization policies, compliance requirements, and recipes baked into them.

Chapter 5, Exploring Infrastructure Platform Patterns, looks at how the success of running an infrastructure platform product within an organization requires a few key patterns that we can use with Crossplane. This chapter will explore these patterns in detail.

Chapter 6, More Crossplane Patterns, explores more Crossplane patterns that are useful for our day-to-day work. We will learn about most of these patterns with a hands-on journey.

Chapter 7, Extending and Scaling Crossplane, covers two unique aspects that make Crossplane extendable and scalable. The first part will deep dive into the Crossplane providers, and the second part will cover how Crossplane can work in a multi-tenant ecosystem.

Chapter 8, Knowing the Trade-Offs, discusses how managing configuration has many nuances to it. Understanding the configuration clock will help us to categorize tools and understand the trade-offs applicable for each category.

Chapter 9, Using Helm, Kustomize, and KubeVela, concentrates on explaining how to use different configuration management tools that are popular today, such as Helm, Kustomize, and KubeVela.

Chapter 10, Onboarding Applications with Crossplane, looks at how infrastructure provisioning and application onboarding involve a few cross-cutting concerns, such as setting up the source code repositories, the continuous integration workflow, and continuous deployment. This chapter will look at ways to approach application, services, and infrastructure automation with Crossplane in a unified way.

Chapter 11, *Driving the Platform Adoption*, explains that many organizations fail with their technology platform projects because they don't apply the needed product development practices and team topology. This chapter aims to help understand the aspects required to build and adopt a successful infrastructure platform.

To get the most out of this book

Please go through the documentation at `https://kubernetes.io/docs/concepts/overview/` to understand the basic concepts. All code examples are tested using the Kind Kubernetes cluster (`https://kind.sigs.k8s.io/ - v1.21.1`) and Crossplane version 1.5.0 as the control plane. However, they should work with future version releases too.

Software/hardware covered in the book	Operating system requirements
Kind Kubernetes cluster – v1.21.1	Windows, macOS, or Linux
Crossplane – v1.5.0 (minimum Kubernetes v1.16.0)	
Helm v3.8.0 (minimum Kubernetes v3.0.0)	
GCP Crossplane provider – v0.18.0	
AWS Crossplane provider – v0.23.0	

Note that for Crossplane installation, you should have a minimum Kubernetes version of v1.16.0.

If you are using the digital version of this book, we advise you to type the code yourself or access the code from the book's GitHub repository (a link is available in the next section). Doing so will help you avoid any potential errors related to the copying and pasting of code.

Download the example code files

You can download the example code files for this book from GitHub at `https://github.com/PacktPublishing/End-to-End-Automation-with-Kubernetes-and-Crossplane`. If there's an update to the code, it will be updated in the GitHub repository.

We also have other code bundles from our rich catalog of books and videos available at `https://github.com/PacktPublishing/`. Check them out!

Download the color images

We also provide a PDF file that has color images of the screenshots and diagrams used in this book. You can download it here: `https://packt.link/1j9JK`.

Conventions used

There are a number of text conventions used throughout this book.

`Code in text`: Indicates code words in text, database table names, folder names, filenames, file extensions, pathnames, dummy URLs, user input, and Twitter handles. Here is an example: "Resources such as Pods, Deployments, Jobs, and StatefulSets belong to the `workload` category."

A block of code is set as follows:

```
# List all resources
kubectl api-resources

# List resources in the "apps" API group
kubectl api-resources --api-group=apps

# List resources in the "networking.k8s.io" API group
kubectl api-resources --api-group=networking.k8s.io
```

When we wish to draw your attention to a particular part of a code block, the relevant lines or items are set in bold:

```
apiVersion: "book.imarunrk.com/v1"
kind: "CloudDB"
metadata:
  name: "aws _ RDS"
spec:
  type: "sql"
  cloud : "aws"
```

Any command-line input or output is written as follows:

```
% kubectl get all -n crossplane-system
helm delete crossplane --namespace crossplane-system
```

Bold: Indicates a new term, an important word, or words that you see onscreen. For instance, words in menus or dialog boxes appear in **bold**. Here is an example: "Go to the **IAM** section in the AWS web console and click **Add a user**."

> **Tips or Important Notes**
> Appear like this.

Get in touch

Feedback from our readers is always welcome.

General feedback: If you have questions about any aspect of this book, email us at customercare@packtpub.com and mention the book title in the subject of your message.

Errata: Although we have taken every care to ensure the accuracy of our content, mistakes do happen. If you have found a mistake in this book, we would be grateful if you would report this to us. Please visit www.packtpub.com/support/errata and fill in the form.

Piracy: If you come across any illegal copies of our works in any form on the internet, we would be grateful if you would provide us with the location address or website name. Please contact us at copyright@packt.com with a link to the material.

If you are interested in becoming an author: If there is a topic that you have expertise in and you are interested in either writing or contributing to a book, please visit authors.packtpub.com.

Share Your Thoughts

Once you've read *End-to-End Automation with Kubernetes and Crossplane*, we'd love to hear your thoughts! Scan the QR code below to go straight to the Amazon review page for this book and share your feedback.

https://packt.link/r/1-801-81154-7

Your review is important to us and the tech community and will help us make sure we're delivering excellent quality content.

Part 1: The Kubernetes Disruption

This part of the book will cover the context of why Kubernetes won the war of application deployment automation and how it is evolving into a new trend in infrastructure automation.

This part comprises the following chapters:

1
Introducing the New Operating Model

Many think that Kubernetes won the container orchestration war because of its outstanding ability to manage containers. But Kubernetes is much more than that. In addition to handling container orchestration at scale, Kubernetes introduced a new IT operating model. There is always a trap with anything new. We tend to use a new tool the old way because of our tendencies. Understanding how Kubernetes disrupted IT operations is critical for not falling into these traps and achieving successful adoption. This chapter will dive deep into the significant aspects of the new operating model.

We will cover the following topics in this chapter:

- The Kubernetes journey
- Characteristics of the new operating model
- The next Kubernetes use case

The Kubernetes journey

The Kubernetes journey to become the leading container orchestration platform has seen many fascinating moments. Kubernetes was an open source initiative by a few Google engineers based on an internal project called Borg. From day one, Kubernetes had the advantage of heavy production usage at Google and more than a decade of active development as Borg. Soon, it became more than a small set of Google engineers, with overwhelming community support. The container orchestration war was a tough fight between Docker, Mesosphere DC/OS, Kubernetes, Cloud Foundry, and AWS **Elastic Container Service (ECS)** from 2015. Kubernetes was outperforming its peers slowly and steadily.

Initially, Docker, Mesosphere, and Cloud Foundry announced native support for Kubernetes. Finally, in 2017, AWS announced ECS for Kubernetes. Eventually, all the cloud providers came up with a managed Kubernetes offering. The rivals had no choice other than to provide native support for Kubernetes because of its efficacy and adoption. These were the winning moments for Kubernetes in the container orchestration war. Furthermore, it continued to grow to become the core of the cloud-native ecosystem, with many tools and patterns evolving around it. The following diagram illustrates the container orchestration war:

Figure 1.1 – The container orchestration war

Next, let's learn about the characteristics of the new operating model.

Characteristics of the new operating model

Understanding how Kubernetes can positively impact IT operations will provide a solid base for the efficient adoption of DevOps in application and infrastructure automation. The following are some of the significant characteristics of the Kubernetes operating model:

- Team collaboration and workflows
- Control theory
- Interoperability
- Extensibility
- New architecture focus
- Open source, community, and governance

Let's look at these characteristics in detail in the following sections.

> **Important Note**
>
> Before we dive deep, it's critical to understand that you are expected to have a basic prior understanding of Kubernetes architecture and its building block resources, such as Pods, Deployments, Services, and namespaces. New to Kubernetes? Looking for a guide to understand the basic concepts? Please go through the documentation at `https://kubernetes.io/docs/concepts/overview/`.

Team collaboration and workflows

All Kubernetes resources, such as Pods, volumes, Services, Deployments, and Secrets are persistent entities stored in `etcd`. Kubernetes has well-modeled RESTful APIs to perform CRUD operations over these resources. The Create, Update, and Deletion operations to the `etcd` persistence store is a state change request. The state change is realized asynchronously with the Kubernetes control plane. There are a couple of characteristics of these Kubernetes APIs that are very useful for efficient team collaboration and workflows:

- Declarative configuration management
- Multi-persona collaboration

Declarative configuration management

We express our automation intent to the Kubernetes API as data points, known as the **record of intent**. The record does not carry any information about the steps to achieve the intention. This model enables a pure declarative configuration to automate workloads. It is easier to manage automation configuration as data points in Git than code. Also, expressing the automation intension as data is less prone to bugs, and easy to read and maintain. Provided we have a clear Git history, a simple intent expression, and release management, collaboration over the configuration is easy. The following is a simple record of intent for an NGINX Pod deployment:

```
apiVersion: v1_
kind: Pod
metadata:
  name: proxy
spec:
  containers:
    - name: proxy-image
      image: Nginx
      ports:
        - name: proxy-port
          containerPort: 80
          protocol: TCP
```

Even though many new-age automation tools are primarily declarative, they are weak in collaboration because of missing well-modeled RESTful APIs. The following multi-persona collaboration section will discuss this aspect more. The combination of declarative configuration and multi-persona collaboration makes Kubernetes a unique proposition.

Multi-persona collaboration

With Kubernetes or other automation tools, we abstract the data center fully into a single window. Kubernetes has a separate API mapping to each infrastructure concern, unlike other automation tools. Kubernetes groups these concerns under the construct called API groups, of which there are around 20. API groups break the monolith infrastructure resources into minor responsibilities, providing segregation for different personas to operate an infrastructure based on responsibility. To simplify, we can logically divide the APIs into five sections:

* **Workloads** are objects that can help us to manage and run containers in the Kubernetes cluster. Resources such as Pods, Deployments, Jobs, and StatefulSets belong to the `workload` category. These resources mainly come under the `apps` and `core` API groups.

* **Discovery and load balancers** is a set of resources that helps us stitch workloads with load balancers. People responsible for traffic management can have access to these sets of APIs. Resources such as Services, NetworkPolicy, and Ingress appear under this category. They fall under the `core` and `networking.k8s.io` API groups.

* **Config and storage** are resources helpful to manage initialization and dependencies for our workloads, such as ConfigMaps, Secrets, and volumes. They fall under the `core` and `storage.k8s.io` API groups. The application operators can have access to these APIs.

* **Cluster resources** help us to manage the Kubernetes cluster configuration itself. Resources such as Nodes, Roles, `RoleBinding`, `CertificateSigningCertificate`, `ServiceAccount`, and namespaces fall under this category, and cluster operators should access these APIs. These resources come under many API groups, such as `core`, `rbac`, `rbac.authorization.k8s.io`, and `certificates.k8s.io`.

* **Metadata** resources are helpful to specify the behavior of a workload and other resources within the cluster. A HorizontalPodAutoScaler is a typical example of metadata resources defining workload behavior under different load conditions. These resources can fall under the `core`, `autoscaling`, and `policy` API groups. People responsible for application policies or automating architecture characteristics can access these APIs.

Note that the core API group holds resources from all the preceding categories. Explore all the Kubernetes resources yourself with the help of the `kubectl` comments. A few comment examples are as follows:

```
# List all resources
kubectl api-resources


# List resources in the "apps" API group
kubectl api-resources --api-group=apps


# List resources in the "networking.k8s.io" API group
kubectl api-resources --api-group=networking.k8s.io
```

The following screenshots give you a quick glimpse of resources under the `apps` and `networking.k8s.io` API groups, but I would highly recommend playing around to look at all resources and their API groups:

```
arunramakani@Aruns-MacBook-Pro ~ % kubectl api-resources --api-group=apps
NAME                     SHORTNAMES      APIVERSION      NAMESPACED      KIND
controllerrevisions                      apps/v1         true            ControllerRevision
daemonsets               ds              apps/v1         true            DaemonSet
deployments              deploy          apps/v1         true            Deployment
replicasets              rs              apps/v1         true            ReplicaSet
statefulsets             sts             apps/v1         true            StatefulSet
```

Figure 1.2 – Resources under the apps API group

The following are the resources under the `network.k8s.io` API group:

```
arunramakani@Aruns-MacBook-Pro ~ % kubectl api-resources --api-group=networking.k8s.io
NAME                SHORTNAMES    APIVERSION            NAMESPACED    KIND
ingressclasses                    networking.k8s.io/v1  false         IngressClass
ingresses           ing           networking.k8s.io/v1  true          Ingress
networkpolicies     netpol        networking.k8s.io/v1  true          NetworkPolicy
```

Figure 1.3 – Resources under the network.k8s.io API group

We can assign RBAC for teams based on individual resources or API groups. The following diagram represents the developers, application operators, and cluster operators collaborating over different concerns:

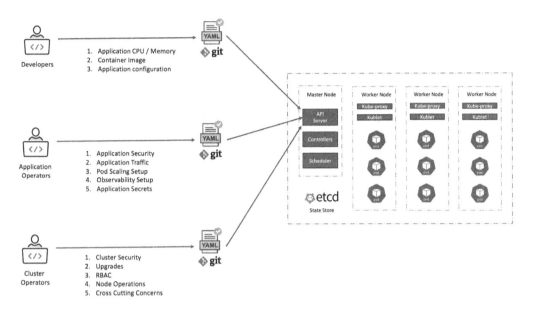

Figure 1.4 – Team collaboration

This representation may vary for you, based on an organization's structure, roles, and responsibilities. Traditional automation tools are template-based, and it's difficult for teams to collaborate. It leads to situations where policies are determined and implemented by two different teams. Kubernetes changed this operating model by enabling different personas to collaborate directly by bringing down the friction in collaboration.

Control theory

Control theory is a concept from engineering and mathematics, where we maintain the desired state in a dynamic system. The state of a dynamic system changes over time with the environmental changes. Control theory executes a continuous feedback loop to observe the output state, calculate the divergence, and then control input to maintain the system's desired state. Many engineering systems around us work using control theory. An air conditioning system with a continuous feedback loop to maintain temperature is a typical example. The following illustration provides a simplistic view of control theory flow:

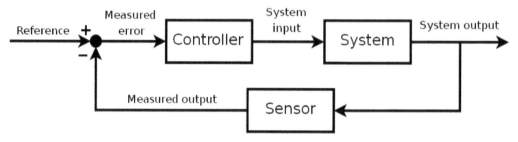

Figure 1.5 – Control theory flow

Kubernetes has a state-of-the-art implementation of control theory. We submit our intention of the application's desired state to the API. The rest of the automation flow is handled by Kubernetes, marking an end to the human workflow once the API is submitted. Kubernetes controllers run a continuous reconciliation loop asynchronously to ensure that the desired state is maintained across all Kubernetes resources, such as Pods, Nodes, Services, Deployments, and Jobs. The controllers are the central brain of Kubernetes, with a collection of controllers responsible for managing different Kubernetes resources. Observe, analyze, and react are the three main functions of an individual controller:

- **Observe**: Events relevant to the controller's resources are received by the observer. For example, a deployment controller will receive all the deployment resource's `create`, `delete`, and `update` events.

- **Analyze**: Once the observer receives the event, the analyzer jumps in to compare the current and desired state to find the delta.

- **React**: Performs the needed action to bring the resources back into the desired state.

The control theory implementation in Kubernetes changed the way IT performs in day one and day two operations. Once we express our intention as data points, the human workflow is over. The machine takes over the operations in asynchronous mode. Drift management is no longer part of the human workflow. In addition to the existing controllers, we can extend Kubernetes with new controllers. We can easily encode any operational knowledge required to manage our workload into a custom controller (operators) and hand over the custom day two operations to machines:

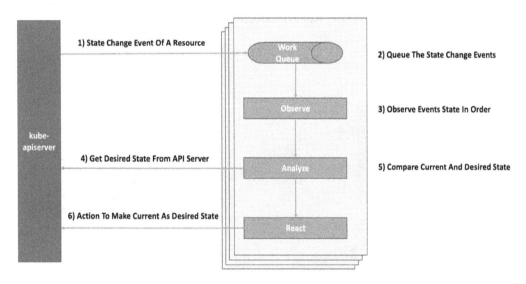

Figure 1.6 – The Kubernetes controller flow

Interoperability

The Kubernetes API is more than just an interface for our interaction with the cluster. It is the glue holding all the pieces together. kubectl, the schedulers, kubelet, and the controllers create and maintain resources with the help of kube-apiserver. kube-apiserver is the only component that talks to the etcd state store. kube-apiserver implements a well-defined API interface, providing state observability from any Kubernetes component and outside the cluster. This architecture of kube-apiserver makes it interoperable with the ecosystem. Other infrastructure automation tools such as Terraform, Ansible, and Puppet do not have a well-defined API to observe the state.

Take observability as an example. Many observability tools evolved around Kubernetes because of the interoperable characteristic of `kube-apiserver`. For contemporary digital organizations, continuous observability of state and a feedback loop based on it is critical. End-to-end visibility in the infrastructure and applications from the perspective of different stakeholders provides a way to realize operational excellence. Another example of interoperability is using various configuration management tools, such as Helm as an alternative to `kubectl`. As the *record of intent* is pure YAML or JSON data points, we can easily interchange one tool with another. The following diagram provides a view of `kube-apiserver` interactions with other Kubernetes components:

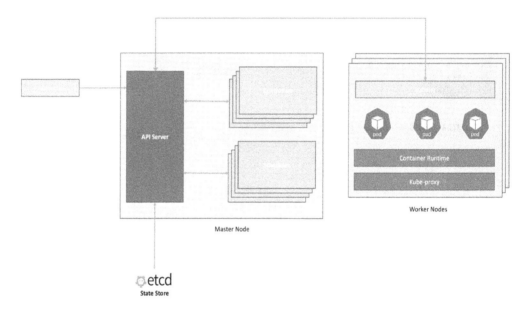

Figure 1.7 – Kubernetes API interactions

Interoperability means many things to IT operations. Some of the benefits are as follows:

- Easy co-existence with the organization ecosystem.

- Kubernetes itself will evolve and be around for longer.

- Leveraging an existing skill set by choosing known ecosystem tools. For example, we can use Terraform for Kubernetes configuration management to take advantage of a team's knowledge in Terraform.

- Hypothetically keeping the option open for migrating away from Kubernetes in the future. (Kubernetes APIs are highly modular, and we can interchange the underlying components easily. Also, a pure declarative config is easy to migrate away from Kubernetes if required.)

Extensibility

Kubernetes' ability to add new functionalities is remarkable. We can look at the extensibility in three different ways:

- Augmenting Kubernetes core components
- Interchangeability of components
- Adding new resource types

Augmented Kubernetes core components

This extending model will either add additional functionality to the core components or alter core component functionality. We will look at a few examples of these extensions:

- **kubectl plugins** are a way to attach sub-commands to the `kubectl` CLI. They are executables added to an operator's computer in a specific format without changing the `kubectl` source in any form. These extensions can combine a process that takes several steps into a single sub-command to increase productivity.

- **Custom schedulers** are a concept that allows us to modify Kubernetes' resource scheduling behavior. We can even register multiple schedulers to run parallel to each other and configure them for different workloads. The default scheduler can cover most of the general use cases. Custom schedulers are needed if we have a workload with a unique scheduling behavior not available in the default scheduler.

- **Infrastructure plugins** are concepts that help to extend underlying hardware. The device, storage, and network are the three different infrastructure plugins. Let's say a device supports GPU processing – we require a mechanism to advertise the GPU usage details to schedule workload based on GPU.

Interchangeability of components

The interoperability characteristics of Kubernetes provide the ability to interchange one core component with another. These types of extensions bring new capabilities to Kubernetes. For example, let's pick up the virtual `kubelet` project (`https://github.com/virtual-kubelet/virtual-kubelet`). `Kubelet` is the interface between the Kubernetes control plane and the virtual machine nodes where the workloads are scheduled. `Virtual kubelet` mimics a node in the Kubernetes cluster to enable resource management with infrastructure other than a virtual machine node such as Azure Container Instances or AWS Fargate. Replacing the Docker runtime with another container runtime environment such as Rocket is another example of interchangeability.

Adding new resource types

We can expand the scope of the Kubernetes API and controller to create a new custom resource, also known as **CustomResourceDefinition** (**CRD**). It is one of the powerful constructs used for extending Kubernetes to manage resources other than containers. Crossplane, a platform for cloud resource management, falls under this category, which we will dive deep into in the upcoming chapters. Another use case is to automate our custom IT day one and day two processes, also known as the operator pattern. For example, tasks such as deploying, upgrading, and responding to failure can be encoded into a new Kubernetes operator.

People call Kubernetes a platform to build platforms because of its extensive extendibility. They generally support new use cases or make Kubernetes fit into a specific ecosystem. Kubernetes presents itself to IT operations as a universal abstraction by extending and supporting every complex deployment environment.

Architecture focus

One of the focuses of architecture work is to make the application deployment architecture robust to various conditions such as virtual machine failures, data center failures, and diverse traffic conditions. Also, resource utilization should be optimum without any wastage of cost in over-provisioned infrastructure. Kubernetes makes it simple and unifies how to achieve architecture characteristics such as reliability, scalability, availability, efficiency, and elasticity. It relieves architects from focusing on infrastructure. Architects can now focus on building the required characters into the application, as achieving them at the infrastructure level is not complex anymore. It is a significant shift in the way traditional IT operates. Designing for failure, observability, and chaos engineering practices are becoming more popular as areas for architects to concentrate on in the world of containers.

Portability is another architecture characteristic Kubernetes provides to workloads. Container workloads are generally portable, but dependencies are not. We tend to introduce dependencies with other cloud components. Building portability into application dependencies is another architecture trend in recent times. It's visible with the 2021 InfoQ architecture trends (`https://www.infoq.com/articles/architecture-trends-2021/`). In the trend chart, design for portability, Dapar, the Open Application Model, and design for sustainability are some of the trends relevant to workload portability. We are slowly moving in the direction of portable cloud providers.

With the deployment of workloads into Kubernetes, our focus on architecture in the new IT organization has changed forever.

Open source, community, and governance

Kubernetes almost relieves people from working with machines. Investing in such a high-level abstraction requires caution, and we will see whether the change will be long-lasting. Any high-level abstraction becoming a meaningful and long-lasting change requires a few characteristics. Being backed by almost all major cloud providers, Kubernetes has those characteristics. The following are the characteristics that make Kubernetes widely accepted and adopted.

Project ownership

Project ownership is critical for an open source project to succeed and drive universal adoption. A widely accepted foundation should manage open source projects rather than being dominated by an individual company, and the working group driving the future direction should have representations from a wide range of companies. It will reflect the neutrality of the project, where every stakeholder can participate and benefit from the initiative. Kubernetes fits very well into this definition. Even though Kubernetes originated from a project by a few Google engineers, it soon became part of the **Cloud Native Computing Foundation** (**CNCF**). If we look at the governing board and members of the CNCF, we can see that there is representation from all top technology firms (`https://www.cncf.io/people/governing-board/` & `https://www.cncf.io/about/members/`). Kubernetes also has special interest groups and working groups and is also represented by many technology companies, including all cloud providers.

Contribution

Kubernetes is one of the high-velocity projects in GitHub, with more than 3,000 contributors. With a high velocity of commits from the community, Kubernetes looks sustainable. Also, there is a high volume of documentation, books, and tutorials available. Above all, we have too many ecosystem tools and platforms evolving around Kubernetes. It makes developing and deploying workloads on Kubernetes easier.

Open standards

As the scope of Kubernetes abstraction is not tiny, it did not attempt to solve all the problems by itself. Instead, it depended on a few open standards to integrate existing widely accepted tools. It also encouraged the ecosystem to develop new tools aligning to open standards. For example, Kubernetes can work with any container runtimes such as Docker and Rocker, which comply with the standard **Container Runtime Interface** (**CRI**). Similarly, any networking solution that complies with the **Container Networking Interface** (**CNI**) can be a networking solution for Kubernetes.

Kubernetes' method of open source governance provides a few advantages to IT operations:

- Kubernetes is sustainable and organizations can invest confidently.
- Wider adoption will maintain a strong talent pool.
- Strong community support.

The preceding section concludes the critical aspects of the new Kubernetes IT operating model. While we have looked at the benefits of every individual characteristic, we also have advantages when we combine them. For example, platforms such as Crossplane are evolving by taking advantage of the multiple aspects discussed previously.

The next Kubernetes use case

In the last few years, many organizations have taken advantage of the disruptive application deployment operating model provided by Kubernetes. The pattern of segregating the intent expression with data points and then a control plane taking over the rest of the automation is known as **Infrastructure as Data** (**IaD**), a term coined by Kelsey Hightower. Many from the Kubernetes community believe that containers are only the first use case for this pattern, and many more will follow in the coming years. A new use case is evolving, with the launch of Crossplane in late 2018 seen as the next big use case for Kubernetes. Crossplane brings the goodness of the Kubernetes operating model to the world of cloud infrastructure provisioning and management. This trend will see people move away from traditional **Infrastructure as Code** (**IaC**), using tools such as Terraform and Ansible, to IaD with Crossplane and Kubernetes. This move addresses the current limitations with the IaC model and unifies the approach of automating applications and infrastructure.

Summary

Kubernetes offers many new aspects to the IT operating model, aligned with modern digital organization expectations. Understanding how Kubernetes disrupts the day one and day two IT operations is key to its successful adoption. This chapter covered the details of the new operating model provided by Kubernetes. In the next chapter, we will look at the limitations of IaC for cloud infrastructure management and introduce Kubernetes control plane-based infrastructure management as the new-age alternative.

2

Examining the State of Infrastructure Automation

This chapter will look at the history of infrastructure automation, its evolution, and its current state. We will explore how the evolving situation in the cloud-native ecosystem and agile engineering practices exposes the limitations of **Infrastructure as Code** (**IaC**). We will also examine how control plane-based infrastructure automation is a cutting-edge technique that solves the limitations of IaC and can change the DevOps operating model to move software engineering further in a positive direction.

The chapter will dive deep into the following topics:

- The history of infrastructure automation
- The limitations of IaC
- The need for end-to-end automation
- Multi-cloud automation requirements
- Crossplane as a cloud control plane
- Other similar projects

The history of infrastructure automation

The hardware purchase cycle was the critical factor influencing an organization's infrastructure landscape changes during the 1990s. Back then, there was not much emphasis on infrastructure automation. The time spent from receiving an order to a physical infrastructure becoming available was much more than the effort spent in infrastructure setup. Individual infrastructure engineers and small teams automated repetitive scripting tasks without much industry-wide adaptation. Tools such as CFEngine, launched in 1993 for automating infrastructure configuration, did not have enough adoption during that decade. There was no industry-wide trend to invest in automation because of its minimal benefits and return on investment. In the 2000s, the idea of infrastructure automation slowly got traction because of the following:

- Virtualization techniques
- The cloud

Virtualization brought in the ability to have software representation of resources such as memory, CPU, storage, and network using a hypervisor installed over physical hardware. It brought us into the era of virtual machines, where machines are abstracted away from the underlying physical hardware. We could have multiple virtual machines over single hardware. It gave us many advantages, such as lower costs, minimal downtime, and effective utilization of resources. But the critical advantage was agility in infrastructure engineering, breaking the traditional hardware purchasing cycles. While virtualization was there before the 2000s for a long time, it saw wide adoption much later because of cloud computing.

Different cloud platforms were launched during the late 2000s, adding more agility. We got into the **Infrastructure as a Service** (**IaaS**) era. As we increased our velocity of spinning new virtual machines, new problems evolved. The number of servers to manage was rapidly growing. Also, virtual machines are transient, and we needed to move, modify, and rebuild them quickly. Keeping configurations up to date with the preceding scenarios is challenging. We ended up with snowflake servers because of an error-prone, intensive human effort to manage the virtual machines manually. These limitations made us move toward the widespread adoption of infrastructure automation. New tools such as Puppet, Ansible, Chef, and Terraform quickly evolved, introducing IaC to manage configuration and provisioning of infrastructure the same way as code. Our ability to be agile in infrastructure life cycle management and store the relevant code in Git is the foundation for modern infrastructure engineering. IaC and IaaS is a deadly combination that provides unique characteristics for infrastructure engineering. We made consistent, repeatable, interchangeable, and elastic infrastructure provisioning and configuration management.

The following diagram summarizes the evolution from scripting to IaC:

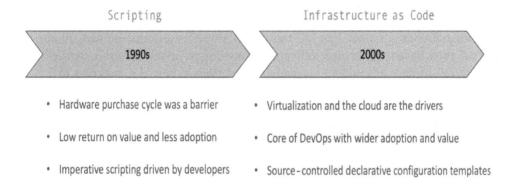

Figure 2.1 – Infrastructure automation evolution

The need for the next evolution

The cloud became the holy grail of infrastructure as we progressed further. Tools such as Terraform, Pulumi, AWS CloudFormation, Google Cloud Deployment Manager, and Azure Resource Manager became the center of IaC. While these tools did well to fulfill their promises, we can see that the next evolution of infrastructure automation is beginning to show up already. Before looking at the next phase of infrastructure automation, it's essential to understand why we need to evolve our tools and practices around infrastructure automation. A few recent trends in the software industry are triggering the next phase of evolution. These trends are the following:

- The limitations of IaC
- The Kubernetes operating model for automation
- Multi-cloud automation requirements

Let's look at each of these trends to justify the need for progression toward the next phase of infrastructure automation.

The limitations of IaC

Most of the widely used infrastructure automation tools are template-based, such as Terraform, Ansible, and Azure Resource Manager. They do not scale well from multiple points of view. It does not mean that IaC tools are not best for automation with all due respect. IaC tools have transformed software engineering positively for more than a decade. We will attempt to explain how evolving situations expose the weakness of template-based IaC tools and how control plane-based tools can be an alternative and the next evolutionary step. Let's pick up Terraform, one of the most popular template-based tools, and look at the limitations. The following are the different limitation issues with Terraform:

- Missing self-service

- A lack of access control

- Parametrization pitfalls

> **Terminology**
>
> The amount of knowledge to be possessed and processed to perform a task is called **cognitive load**. You will come across the term **high team cognitive load** in the upcoming sections, which means that a team must stretch its capacity to hold more knowledge than it usually does to perform day-to-day functions.

Missing self-service

With Terraform, we have too many templates abstracting thousands of cloud APIs. Remembering the usage of each parameter in thousands of templates is not an easy job. Also, infrastructure usage policies come from different teams in an organization, such as security, compliance, product, and architecture. Implementing Terraform automation involves a significant team cognitive load and centralized policy requirements. Hence, many organizations prefer to implement infrastructure automation with centralized platform teams to avoid increased cognitive load on the product team and enable centralized policy management. But template-based automation tools do not support APIs, the best way to provide platform self-service. So, we must build Terraform modules/libraries to create artificial team boundaries and achieve self-service. Modules/libraries are a weak alternative to APIs. They have a couple of problems in enabling platform self-service:

- There is a leak in cognitive load abstraction by the platform team, as using Terraform modules/libraries by the product team means learning Terraform fundamentals at least.

- The team dependencies as modules and libraries require a collaborative model of interaction between the product and platform teams rather than a self-service model. It is against the modern platform topologies, hindering the agility of both platform and product teams.

Alternatively, some organizations outsource the infrastructure provisioning completely to the platform team. The complete centralization hinders the product team's agility, with external coupling for infrastructure provisioning. Few organizations even attempt to decentralize the infrastructure management into the product teams. A complete decentralization will increase team cognitive load and the difficulty of aligning centralized policies across teams. The new evolution needs to find the middle ground with correctly abstracted self-service APIs.

Lack of access control

As we saw in the previous section, building and using Terraform modules requires collaboration between multiple teams. We have access control issues by sharing Terraform modules with product teams for infrastructure provisioning and management. We cannot have precise **Role Based Access Control** (**RBAC**) on individual resources required by the product team, and we will leak the underlying cloud credentials with all the necessary permissions required by the modules. For example, a Terraform module to provision Cosmos DB requires Azure credentials for database and **Virtual Private Cloud** (**VPC**) provisioning. But the access needed for the product team is only to create the database, and they don't need to modify the VPC directly. In addition to this, we also have version management issues with modules/libraries. It requires a coordinated effort between all product teams, creating friction on a module/library's version upgrade. A highly interoperable API-based infrastructure automation abstraction can solve collaboration and access control issues.

Parameterization pitfalls

Parameterization pitfalls are one of the general issues with any template-based solution, be it an infrastructure automation tool or otherwise. We create parameter placeholders for variables with changing values in any template-based solution. These solutions are easy to implement, understand, and maintain. They work well if we are operating at a small scale. When we try to scale template-based solutions, we end up with either one of the following issues:

- As time passes, we will have requirements to parameterize new variables, and slowly, we will expose all the variables at some point in time. It will erode the abstraction we created using templates. Looking at any Helm chart will show this clearly, where almost everything is a parameter.

- We may decide to fork the main template to implement customization for a specific use case. Forks are challenging to keep up to date, and as the number of forks increases, it will be challenging to maintain the policies across the templates.

Parameterization is generally not a perfect abstraction when we operate at scale.

> **Important Note**
> Parameterization pitfalls is a critical topic to understand in detail for DevOps engineers. In a later chapter, we will look at the configuration clock, a concept of eroding template abstractions as time passes.

A Kubernetes operating model for automation

As we saw in the previous chapter, the control theory implementation of Kubernetes entirely changed the IT operations around application automation. Infrastructure automation as well deserves the same benefits. But traditional infrastructure automation tools lack these attributes, as they don't have an intact control theory implementation. Some of the missing features are the following:

- **Synchronous provisioning** is a crucial scalability issue with Terraform or similar automation tools. The resources are provisioned in a sequence, as described in the dependencies with conventional automation tools. If infrastructure *A* depends on infrastructure *B*, we must respect it while defining the order of execution, and if one of the executions fails, the whole automation fails. The monolithic representation of infrastructure is the key concern here. With Terraform, the monolithic state file is the model representing infrastructure resources. Kubernetes-based automation can change this equation. There will be a continuous reconciliation to move the current state toward the expected state. Hence, we can efficiently manage the dependencies with no order of execution. Infrastructure *A* provisioning may fail initially, but continuous reconciliation will eventually fix the state once infrastructure *B* is available.

- **Modeling team boundaries** is another missing piece with traditional tools. The monolithic Terraform state file is not flexible to model different team boundaries. In the Kubernetes-based automation model, we have resources represented as individual APIs that can be grouped and composed as required by any team structure. We don't need to collect all pieces of automation into a single monolithic model.

- **Drift management** is the process of maintaining the infrastructure in the intended state by protecting it against any unintended or unauthorized changes. Changing the **Identity and Access Management** (**IAM**) policy directly in the cloud console without changing the relevant automation code is an example of drift. Drift management is all about bringing it back to the authorized state. Drift management is impossible with no control plane continuously monitoring the infrastructure's condition and performing reconciliation against the last-executed code. Achieving drift management with an additional external tool will add complexity and not solve all the issues.

- **Automating day 2 concerns** in a standard way is another missing piece with conventional tools. A Kubernetes-based automation model can provide configuration models to support day 2 concerns such as scaling, monitoring, and logging. Also, we can use standard extension points (operators) to automate any custom day 2 problems.

These are a few essential perspectives on what Kubernetes-based infrastructure automation can bring to the table.

Multi-cloud automation requirements

Almost all organizations of a significant size run their workloads in more than one cloud provider. There can be many reasons why an organization is determined to build its infrastructure supported by multiple cloud providers. We will not get into details of these factors, but we must understand the impact of multi-cloud on infrastructure management. Typically, a cloud provider offers managed services, be it basic IaaS such as Amazon EC2 or more abstracted platforms such as AWS Lambda. From the perspective of cloud infrastructure consumers, infrastructure automation is all about the provisioning and life cycle management of these managed services in an automated fashion after applying all the in-house policies. Organizations use infrastructure automation tools to build an abstraction over the cloud infrastructure APIs to encode all the in-house policies.

To support multi-cloud capability requires a lot of work, as it brings in new requirements. Think about the multi-cloud environment. Embedding policies into the automation scripts of every cloud provider is a hell of a lot of work. Even if we do that after making a significant effort, keeping these policies in sync across the automation scripts involves friction and is error-prone. A centralized experience in authentication, authorization, billing, monitoring, and logging across cloud providers will be an added advantage for an organization to provide a unified experience. Achieving these cross-cutting concerns with traditional automation tools requires a lot of custom engineering, making our platform team big. What we need is a centralized control plane, abstracting cross-cutting concerns and policies.

The following figure represents how an API-driven centralized control plane can provide a unified experience for product teams:

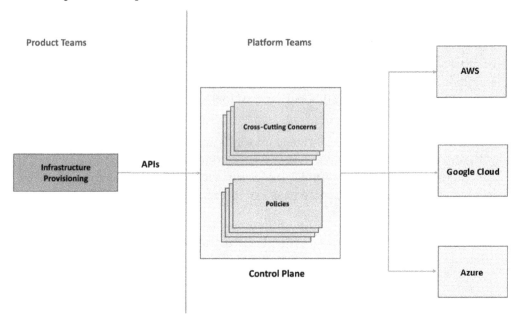

Figure 2.2 – A multi-cloud control plane

Tools such as Terraform or Pulumi help with these problems to some extent, but they are not end-to-end automation, have scalability issues, and require custom engineering to build on. Also, these tools are not unbiased open source projects. The companies who initially created these open source projects and provided enterprise offerings dominate the control of the former. Now that we are all convinced that the next evolution of infrastructure automation is required, it's time to define the attributes needed by such tools. The subsequent development of infrastructure automation should be a control plane-based, fully community-driven solution, powered by APIs. The following diagram summarizes the evolution from **Infrastructure as Code-** to **Central Control Plane**-based automation:

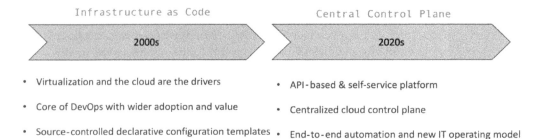

Figure 2.3 – The next evolution

Crossplane as a cloud control plane

Crossplane, a modern control plane-based infrastructure automation platform built on Kubernetes, matches all the attributes required for the next evolution of infrastructure engineering. With Crossplane, we can assemble infrastructure from multiple cloud providers to expose them as a high-level API. These APIs can provide a universal experience across teams, irrespective of the underlying cloud vendor. While composing the APIs for the product team, the platform team can use different resource granularity to suit the organization's structure. Such carefully crafted APIs for infrastructure automation will facilitate self-service, multi-persona collaboration with precise RBAC, less cognitive load, continuous drift management, and dependency management with asynchronous reconciliation. Above all, the platform team can compose these APIs in a no-code way with configurations. Finally, we can have a lean platform team, as highly recommended by modern team topologies.

Crossplane is nothing but a set of custom controllers that extends Kubernetes for managing infrastructure from different vendors. Being built on Kubernetes as a new API extension, Crossplane inherits all the goodness of the Kubernetes operating model and can leverage the rich ecosystem of cloud-native tools. Additionally, this can unify the way we automate applications and infrastructure. Crossplane can cover end-to-end automation of both day 1 and day 2 concerns. Infrastructure provisioning, encoding policies, governance, and security constraints are the day 1 concern we can automate. We can cover drift management, upgrades, monitoring, and scaling on day 2. Above all, it follows the Kubernetes model of open source governance through the **Cloud Native Computing Foundation** (**CNCF**). The following figure represents how Crossplane works with Kubernetes:

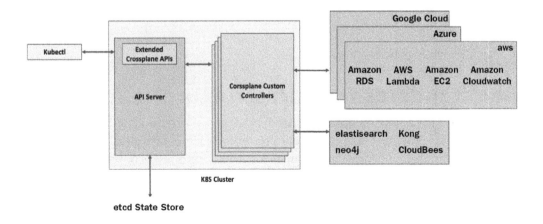

Figure 2.4 – The Crossplane control plane

To adopt platforms as a universal control plane requires a much closer look at open source governance and ecosystem acceptance. The following sections will look deep into these aspects.

A universal control plane

Launched in 2018 as an open source project, Crossplane took steps to become a universally accepted control plane. The project's donation to CNCF in 2020 was the next significant step. It helped Crossplane become a foundation-driven, open source initiative with broader participation rather than just becoming another open source project. Initially, it was a sandbox project but did not stop there. In 2021, it was accepted as an incubating project. Above all, Crossplane is simply another extension to Kubernetes, already an accepted platform for application DevOps. It also means that the entire ecosystem of tools available for Kubernetes is also compatible with Crossplane. Teams can work with the existing set of tools without much cognitive load:

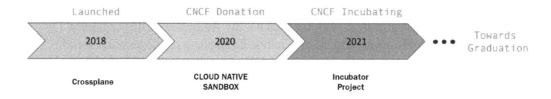

Figure 2.5 – The journey

Crossplane has a few more unique attributes compelling it to be accepted as a universal control plane. The attributes are the following:

- Open standards for infrastructure vendors
- Wider participation
- Cloud provider partnerships

Open standards for infrastructure vendors

Crossplane uses the **Crossplane Resource Model** (**XRM**), an extension of the **Kubernetes Resource Model** (**KRM**), as the open standard for infrastructure providers. It solves issues such as naming identity, package management, and inter-resource references when infrastructure offerings from different vendors are consolidated into a single control plane. The Crossplane community has developed these standards to enforce how infrastructure providers can integrate into the centralized Crossplane control plane. The ability to compose different infrastructures in a uniform and no-code way has its foundation on this standardization.

Wider participation

Upbound was the company that initially created Crossplane. They provide enterprise offerings for organizations that require support and additional services. But to become a universal control plane, Upbound cannot be the only enterprise Crossplane provider. Any vendor should be able to make an enterprise offering. With Crossplane gaining CNCF incubation status, a lot of work is happening in this area. CNCF and the Crossplane community have introduced something called the **Crossplane Conformance Program**. It's an initiative run by CNCF (`https://github.com/cncf/crossplane-conformance`). The idea is to create foundation governance for any vendors to pick up Crossplane open source, build additional features, and offer a **CNCF-certified** version. It is very similar to **Kubernetes-certified distribution**, a program run by CNCF where all vendors pick up the base Kubernetes version and offer it as a certified version. The Crossplane Conformance Program works on two levels:

- **Providers**: On one level, infrastructure providers will be interested in building respective Crossplane controllers to enable customers to use their offerings through Crossplane. It requires following the standards set by XRM. CNCF will ensure this happens by certifying the providers built by infrastructure vendors.

- **Distribution**: On another level, many vendors will be interested in providing the Crossplane enterprise offering. The Crossplane Conformance Program enables this support.

Read more about the Crossplane Conformance Program at `https://github.com/cncf/crossplane-conformance/blob/main/instructions.md`.

The cloud provider partnerships

Crossplane has an excellent partnership ecosystem with all major cloud providers. There have been production-ready Crossplane providers for all the major cloud providers for quite some time now. Initially, IBM joined the Crossplane community and released its first version of the provider in 2020. Similarly, AWS and Azure made Crossplane providers part of their code generation pipeline to ensure that the newest provider is available up front for all their cloud resources. Alibaba is experimenting with Crossplane on many of its internal initiatives and also has a production-ready provider. Similarly, there has been a **Google Cloud Platform** (**GCP**) provider managed by the community. These partnerships and community efforts make Crossplane a compelling, widely accepted universal control plane offering.

Other similar projects

A few other Kubernetes-based infrastructure automation projects share common interests and support similar use cases such as Crossplane. These projects extend Kubernetes with APIs and custom controllers identical to Crossplane architecture. This section will look at those tools to have a comprehensive comparison with Crossplane. The following is a list of a few such projects:

- **Kubernetes' Service Catalog** by the open service broker enables life cycle management of cloud resources from Kubernetes. Like Crossplane, it works as a Kubernetes controller extension. But it does not have a solid framework to compose infrastructure recipes with policy guardrails. Also, we can't model the API for different team boundaries. The open service broker Kubernetes Service Catalog is not designed for platform teams to build reusable recipes with encoded policies. Typically, this means that we have to struggle with policy enforcement and a high cognitive load on the teams to understand cloud offerings in detail.

- **AWS Controllers for Kubernetes (ACK)** is a Kubernetes-based extension developed by AWS to manage its resources from the Kubernetes cluster using controllers. Again, it does not have a framework to compose infrastructure recipes and model APIs. Also, this does not work cross-cloud and is meant to be used only with AWS.

- The **GCP Config Connector** is a replacement developed by Google for the GCP service catalog. It works like ACK and inherits identical limitations. An additional point to note is that the GCP Config Connector is not an open source initiative.

None of these tools cover an end-to-end automation use case or provide an ability to compose resources as recipes. We have already seen the limitations of Terraform, AWS CloudFormation, Azure Resource Manager, and similar IaC tools in detail. These were the motivations that the Crossplane creators had when developing such universal abstraction.

Summary

This chapter discussed the details of limitations with IaC. We also looked at why it is inevitable to move toward a control plane automation in the evolving world of software engineering. It brings us to the end of the first part of this book. In summary, part one covered how Kubernetes won the war on application deployment automation and how the same pattern is evolving a new trend in infrastructure automation. The upcoming sections of the book will take us on a hands-on journey to learn Crossplane, Kubernetes configuration management, and ecosystem tools. We also will cover the different nuances and building blocks of developing state-of-the-art cloud infrastructure automation platforms with Crossplane.

In the next chapter, we will learn about automating infrastructure with Crossplane.

Part 2: Building a Modern Infrastructure Platform

This section will cover different nuances of building state-of-the-art infrastructure automation platforms with Kubernetes and Crossplane. It will be a step-by-step journey to learn about Crossplane and create a self-service infrastructure automation platform with the Kubernetes API.

This part comprises the following chapters:

- *Chapter 3, Automating Infrastructure with Crossplane*
- *Chapter 4, Composing Infrastructure with Crossplane*
- *Chapter 5, Exploring Infrastructure Platform Patterns*
- *Chapter 6, More Crossplane Patterns*
- *Chapter 7, Extending and Scaling Crossplane*

3

Automating Infrastructure with Crossplane

It is time to stop being abstract about ideas and deep-dive into details. Starting with this chapter, we will go on a hands-on journey to implement what we have learned and explore different Crossplane concepts at the same time. Trying out the examples given in this book will ensure that we have the ideas and experience to practice modern infrastructure engineering in our day-to-day jobs. This chapter will specifically cover the detailed architecture of Crossplane and its out-of-the-box features.

The following are the topics covered in this chapter:

- Understanding Custom Resource Definitions and custom controllers
- Understanding the Crossplane architecture
- Installing Crossplane
- Installing and configuring providers
- Multiple provider configuration
- An example of POSTGRES provisioning

Understanding Custom Resource Definitions and custom controllers

Understanding the concept of **Custom Resource Definitions** (**CRDs**) and custom controllers in Kubernetes is vital to know how Crossplane works. Before getting into the Crossplane architecture, we will take a quick look at CRDs and custom controllers.

> Terminology
>
> The term **resources** in Kubernetes represents a collection of objects of a similar kind. Pods, Services, Deployments, namespaces, and many more are the in-built object kinds. Each resource has the respective API endpoints at kube-apiserver.

CRDs are the way to extend the in-built resources list. It adds a new resource kind, including a set of API endpoints at kube-apiserver, to operate over the new resource. The term CRD precisely indicates what it does. The new resource added to Kubernetes using a CRD is called a **Custom Resource** (**CR**). Storing and retrieving a structured object defined with a CRD is not helpful unless backed by a custom controller. Custom controllers are our addition to the in-built controllers. They generally operate over a CR to perform specific actions in a control loop for each API event of a given resource. *Chapter 1, Introducing the New Operating Model*, has already covered the concept of a control loop. Refer to the *Control theory* section if you want to brush up on the idea. Note that the custom controllers do not necessarily always work with a CR. They can work with any existing resources to extend its functionality, which is not in the scope of our discussion. The Prometheus operator is one of the most famous and widely used examples of adding a few new CRs and controllers into the Kubernetes cluster to perform monitoring of workloads.

> Terminology
>
> The term **operator** when applied to Prometheus is a term coined by CoreOS. Operators are nothing but a combination of *CRD + a custom controller + application focus*.

Adding a new CRD

While there are a few ways to add CRDs to Kubernetes, let's add CRDs by creating a `yaml` file and applying the same on the cluster. The `yaml` file provides a structure to the CR. The following are the key attributes of a CRD YAML:

- **ApiVersion** for a CRD YAML falls under `apiextensions.k8s.io/v1beta1`.

- **Kind** is the elements that specify that we are representing a resource as YAML. The name designated for the CRD resource is `CustomResourceDefinition`.

- The **metadata name** is the next critical element that requires a standard format. The format is the plural name (the plural name of the CR we define) + . + group (the API group under which we what to classify the CR). In the following YAML sample, we describe it as `clouddbs.book.imarunrk.com`, where `clouddbs` is the plural name and `book.imarunrk.com` is a newly defined API group. The term plural name is nothing but a representation of the resource name in the plural format. For example, the plural format of the existing resource pod is pods. We can always use that with kubectl commands (`kubectl get pods`).

- **Spec** is the section under which we define the details of the CR. The attributes of this section include the API group, version, scope, and name. This section also describes the validation requirements of the API itself, such as the parameter list, their data types, and whether they are a required parameter.

The following is a sample CRD YAML that represents a cloud database resource. It takes a couple of mandatory parameters, the database type, and the cloud provider name:

```
apiVersion: "apiextensions.k8s.io/v1beta1"
kind: "CustomResourceDefinition"
metadata:
  name: "clouddbs.book.imarunrk.com"
spec:
  group: "book.imarunrk.com"
  version: "v1"
  scope: "Namespaced"
  names:
    plural: "clouddbs"
    singular: "clouddb"
    kind: "CloudDB"
  validation:
    openAPIV3Schema:
```

```
    required: ["spec"]
    properties:
      spec:
      required: ["type","cloud"]
      properties:
        type:
           type: "string"
           minimum: 1
        cloud:
           type: "string"
           minimum: 1
```

Executing the following YAML code will enable new rest endpoints at kube-apiserver to create, modify, and delete the CloudDB resource:

```
# Apply the CRD yaml to the Kubernetes cluster
kubectl create -f crd.yaml
```

The next step is to write the custom controller to manage the CR API events. We can write a controller in many different languages using different open source frameworks. Writing a controller is an advanced topic that is covered later in *Chapter 7, Extending and Scaling Crossplane*. Crossplane providers are nothing but controllers to manage external infrastructure resources – generally, cloud providers' managed services. For now, we must remember that controllers implement functions to execute three actions – observe, analyze, and react in a control loop. In the preceding example, the control loop will create, update, and delete the cloud database based on the resource's creation, update, and delete API events.

Working with the CRD

Once we have the CRD and controller in place, we can start creating and managing the cloud database with `kubectl`. It will work very similarly to other in-built resources such as the Pod. The following YAML is an example of creating Amazon RDS:

```
apiVersion: "book.imarunrk.com/v1"
kind: "CloudDB"
metadata:
```

```
  name: "aws_RDS"
spec:
  type: "sql"
  cloud : "aws"
```

Applying the following command will create a new CloudDB resource:

```
# Apply the CloudDB yaml to the Kubernetes cluster
kubectl create -f aws_rds.yaml
```

Note that the preceding YAML will not create an RDS, as we have not developed and deployed a controller. The example is to explain how CRDs and custom controllers work. Now that we have learned about CRDs and custom controllers, it's time to look at the Crossplane architecture in detail.

Understanding the Crossplane architecture

From what we know so far, Crossplane is nothing but a set of Kubernetes custom controllers and CRDs representing external infrastructure resources. If you take a closer look, Crossplane is much more than a combination of CRDs and custom controllers. Crossplane has four key components. The components are as follows:

- Managed resources
- Providers
- Composite resources
- The Crossplane core

Managed resources

A **Managed Resource** (**MR**) connects a CRD and respective custom controller to represent a single external infrastructure resource. MRs are in a one-to-one mapping with infrastructure resources. For example, CloudSQLInstance is an MR representing Google Cloud SQL. The following diagram depicts the MR mapping for Amazon RDS and Google Cloud Storage:

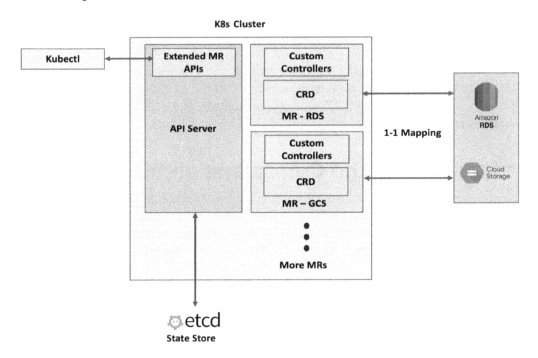

Figure 3.1 – MR mapping

The **Crossplane Resource Model** (**XRM**) is the open standard used when developing an MR. The XRM is an opinionated extension of the **Kubernetes Resource Model** (**KRM**). The XRM sets the standards for external resource naming, dependency management, package management, and so on. MRs are the basic building blocks on which everything else in Crossplane is made. We can use MRs directly to provision an external infrastructure, but this is very rarely done. The best practice is to compose a higher-level API above an MR to consume. We will look at *how* and *why* in a later section of this chapter, as well as a dedicated chapter to address this aspect in detail. The following YAML represents the provisioning of CloudSQLInstance directly using an MR:

```
apiVersion: database.gcp.crossplane.io/v1beta1
kind: CloudSQLInstance
metadata:
```

```
    name: my-GCP-DB
spec:
  forProvider:
    databaseVersion: POSTGRES_9_6
    region: asia-south2
    settings:
      tier: db-n1-standard-1
      dataDiskSizeGb: 10
  writeConnectionSecretToRef:
    namespace: DB
    name: my-GCP-DB-credentials
```

The YAML will provision a GCP Cloud SQL instance with the resource constraints mentioned in the configurations. Since MR is the low-level Crossplane construct mapping to the infrastructure provider API, it will directly support all the attributes available in the infrastructure API. The reconciliation control loop of Crossplane controllers will fill default values assigned by the infrastructure API for the features not provided in the configuration YAML. This concept of MR is called late initialization. Whatever we provide under the forProvider: section will represent the attributes of the infrastructure API. If someone or a parallel process performs an unauthorized change to the infrastructure resource, Crossplane will roll back the changes to the source of truth mentioned in the YAML. The remaining parts of the configuration will help to determine other Crossplane behavior. For example, the preceding YAML has writeConnectionSecretToRef: to decide how to save the database credentials. There can be many more such behavior controls, which we will learn as we go further. The following commands can help look at the details of the created GCP resource and clean up the resources after testing:

```
# View the resources created with

kubectl get cloudsqlinstance my-GCP-DB
kubectl describe cloudsqlinstance my-GCP-DB

# Delete the resources created with

kubectl delete cloudsqlinstance my-GCP-DB
```

We can also import the existing provisioned infrastructure into the Crossplane ecosystem. The MR checks whether the resources named in the configuration YAML already exist before fresh provisioning. When we build that YAML for the existing infrastructure, we can provide the authorized attributes that need to be maintained by the reconciliation loop under `forProvider`. In the next chapter, we will look at an example of importing existing infrastructure into Crossplane.

> **Tip**
> One of the possible Crossplane behavior controls is the deletion policy, specified with an attribute called **deletionPolicy**. It can have two possible values – **Delete**, which is the default, and **Orphan**. While Delete will remove the resource from the infrastructure provider, Orphan will just remove the Crossplane reference.

Providers

Providers are a group of related MRs packaged together as a bundle. Each cloud vendor, other infrastructure services such as Rook (the cloud-native storage for Kubernetes – `https://rook.io/`), and software tools such as Helm have a provider package. The format of these packages is nothing but a container image. Installing a provider can be done either with a configuration YAML or using a Helm package. It requires a ProviderConfig configuration before we can start using these packages. The ProviderConfig resource helps us to set up the infrastructure vendor credentials. We will look at provider installation and ProviderConfig setup in the next section of the chapter.

The following figure represents the AWS and GCP providers extending Crossplane for the respective cloud resource provisioning:

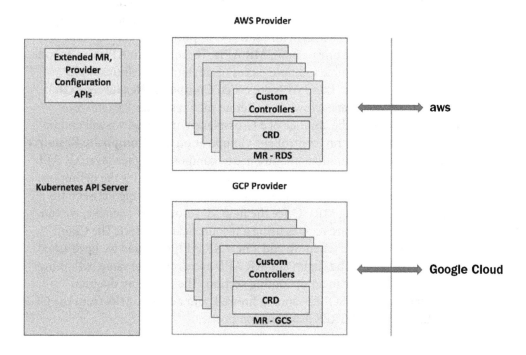

Figure 3.2 – A provider view

Composite resources

Every organization is different in its structure, size, operating model, technology maturity, compliance, and security requirements. All these aspects will directly create a set of rules as policies for infrastructure provisioning and usage. Also, there is a complexity in assembling multiple pieces of the infrastructure stack. Composing is a way to build our custom infrastructure APIs above one or more MR APIs. These custom APIs can encode all the policy guidelines and assemble multiple pieces of infrastructure into simple, easy-to-use infrastructure recipes. These APIs are called **Composite Resources (XRs)**. The critical aspect is that Crossplane allows us to define such resources in a no-code way, just with configurations. In the absence of a Crossplane composite, we will end up building complex Kubernetes custom controllers. Composition and **Composite Resource Definition (XRD)** are the configurations that we use to compose a higher-level XR API. XRD defines the schema of the new custom API that we are building. It's the definition of a new CRD. Composition is the configuration that provides a mapping between the new CRD schema and the existing MRs. Once the new XR resource is available, we can start provisioning the infrastructure recipes using a resource claim object. The Claim API is something that gets created when we add a new XR API, provided we have asked for it in the XRD configuration. The Claim and XR APIs are almost the same with minor differences, which we will explore in an upcoming section. The following diagram represents how we can use CloudSQLInstance, a firewall, and network MRs from the GCP provider to construct an XPOSTGRES database composite:

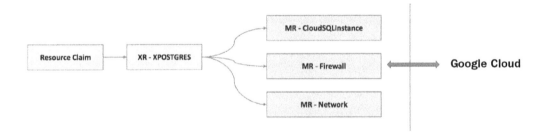

Figure 3.3 – A composite resource

We have thousands of cloud-managed services with many attributes requiring a detailed configuration. It adds a significant cognitive load to shortlist the options and implement the resource provisioning, with good guardrail settings that suit our organization's needs. Every product team in an organization cannot take up this level of cognitive load. Usually, organizations use a platform team to abstract the cognitive load. The composing layer is for platform teams to build such abstractions. Crossplane enables us to expose these abstractions as Kubernetes APIs, allowing self-service for the product teams. The following diagram represents the platform and product team interaction model:

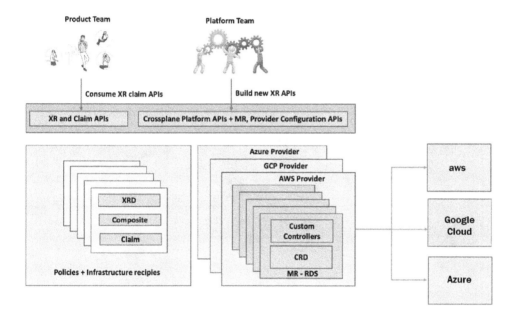

Figure 3.4 – Product and platform team interaction

Crossplane core

Before setting up the providers we are interested in and building XRs above that, we need to have a Crossplane core component installed in the cluster. This component provides the primitives required to manage multiple providers, build new XRs, and build new packages. Again, the core component is a bundle of CRDs and custom controllers. It is the glue that holds everything else about Crossplane together. The following figure represents how all the pieces fit within the Kubernetes ecosystem:

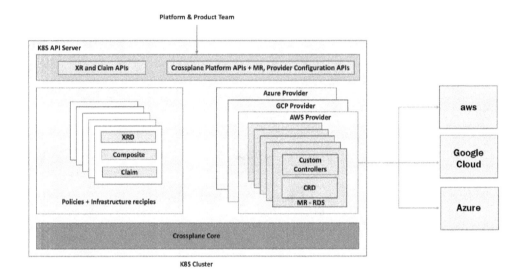

Figure 3.5 – All the pieces of Crossplane

Let's revisit some of the acronyms that we have used so far, which we will use going forward:

- **CRD** stands for **Custom Resource Definition**, a way to extend the Kubernetes API to support new resource types.

- CRs are the resources defined with the CRD. **CR** stands for **Custom Resource**.

- **XRM** stands for **Crossplane Resource Model**, an extension of the **Kubernetes Resource Model**. It is a set of rules set by Crossplane on how to build an XR or MR.

- **MRs** are nothing but **Managed Resources**, a combination of a CRD and custom controllers in a one-to-one mapping with external infrastructure resources.

- **XRD** stands for **Composite Resource Definition**, a definition for building a custom infrastructure API above Crossplane.

- **XR** stands for **Composite Resource**, which represents a custom infrastructure API built with Crossplane.

Installing Crossplane

I have set up a local Kubernetes cluster on my Macintosh computer. We will make this cluster the Crossplane control plane for provisioning resources from Google Cloud Platform in our tutorials. To follow the tutorials, we assume that you already have access to a Kubernetes cluster. If you need help setting up a local Kubernetes cluster, refer to `https://kind.sigs.k8s.io/`. kind is one of the simplest ways to set up a local Kubernetes cluster, but you can work on the tutorials with any Kubernetes cluster setup. The following screenshot gives a quick glimpse at cluster information, versions, and node details:

```
arunramakani@Aruns-MacBook-Pro ~ % kubectl version
Client Version: version.Info{Major:"1", Minor:"21", GitVersion:"v1.21.2", GitCommit:"092fbfbf53427de67cac1e9fa54aaa09a28371d7", GitTreeState:"clean"
, BuildDate:"2021-06-16T12:59:11Z", GoVersion:"go1.16.5", Compiler:"gc", Platform:"darwin/amd64"}
Server Version: version.Info{Major:"1", Minor:"21", GitVersion:"v1.21.1", GitCommit:"5e58841cce77d4bc13713ad2b91fa0d961e69192", GitTreeState:"clean"
, BuildDate:"2021-05-21T23:01:33Z", GoVersion:"go1.16.4", Compiler:"gc", Platform:"linux/amd64"}
arunramakani@Aruns-MacBook-Pro ~ % kubectl get nodes
NAME                STATUS   ROLES                  AGE   VERSION
kind-control-plane  Ready    control-plane,master   84d   v1.21.1
arunramakani@Aruns-MacBook-Pro ~ % kubectl cluster-info
Kubernetes control plane is running at https://127.0.0.1:55598
CoreDNS is running at https://127.0.0.1:55598/api/v1/namespaces/kube-system/services/kube-dns:dns/proxy

To further debug and diagnose cluster problems, use 'kubectl cluster-info dump'.
arunramakani@Aruns-MacBook-Pro ~ %
```

Figure 3.6 – Cluster information, versions, and node details

While there are a few ways to install Crossplane in Kubernetes, we will install it with a Helm chart. Make sure you have Helm installed. Installing Helm is pretty simple on both Macintosh and Windows with the `brew` or `choco` package managers. The following commands can install `helm` in your environment:

```
# Install helm in mac

brew install helm

# Install helm in windows

choco install kubernetes-helm
```

For more installation options, visit `https://helm.sh/docs/intro/install/`. For Crossplane installation, we need a minimum Kubernetes version of v1.16.0 and a minimum Helm version of v3.0.0. Master and stable are the two flavors of Crossplane available. Master has the latest commits, but this version may have stability issues. Stable is a version that is ready for community use and recommended for production usage. We will be using the stable version of the Crossplane in this tutorial. The installation of Crossplane is done in three steps, as follows:

1. Create a new target namespace called `crossplane-system` for installation.

2. Add and update the Crossplane repository to the `helm` repo list.

3. Install Crossplane with `helm` in the target namespace.

The code is as follows:

```
# Step 1: Create target namespace

kubectl create namespace crossplane-system

# Step 2: Add crossplane stable repo to helm and update

helm repo add crossplane-stable \ https://charts.crossplane.io/
stable
helm repo update

# Step 3: Install Crossplane

helm install crossplane --namespace crossplane-system
crossplane-stable/crossplane
```

The Crossplane Helm chart supports quite a few template values for custom configuration options. Replica count is one example, which specifies how many Crossplane Pods are to run for a high-availability setup. Have a look at all possible configuration options available for the Crossplane Helm template at Artifact Hub. The installation screenshot is as follows:

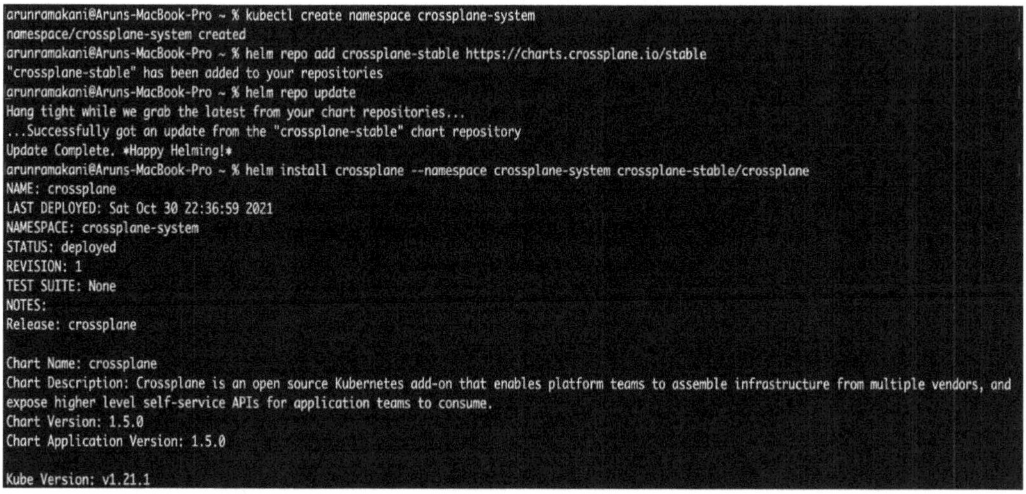

```
arunramakani@Aruns-MacBook-Pro ~ % kubectl create namespace crossplane-system
namespace/crossplane-system created
arunramakani@Aruns-MacBook-Pro ~ % helm repo add crossplane-stable https://charts.crossplane.io/stable
"crossplane-stable" has been added to your repositories
arunramakani@Aruns-MacBook-Pro ~ % helm repo update
Hang tight while we grab the latest from your chart repositories...
...Successfully got an update from the "crossplane-stable" chart repository
Update Complete. ✦Happy Helming!✦
arunramakani@Aruns-MacBook-Pro ~ % helm install crossplane --namespace crossplane-system crossplane-stable/crossplane
NAME: crossplane
LAST DEPLOYED: Sat Oct 30 22:36:59 2021
NAMESPACE: crossplane-system
STATUS: deployed
REVISION: 1
TEST SUITE: None
NOTES:
Release: crossplane

Chart Name: crossplane
Chart Description: Crossplane is an open source Kubernetes add-on that enables platform teams to assemble infrastructure from multiple vendors, and
expose higher level self-service APIs for application teams to consume.
Chart Version: 1.5.0
Chart Application Version: 1.5.0

Kube Version: v1.21.1
```

Figure 3.7 – Crossplane installation

We can remove the Crossplane installation with the standard `helm delete` command:

```
# To remove Crossplane
helm delete crossplane --namespace crossplane-system
```

A screenshot of the result is as follows:

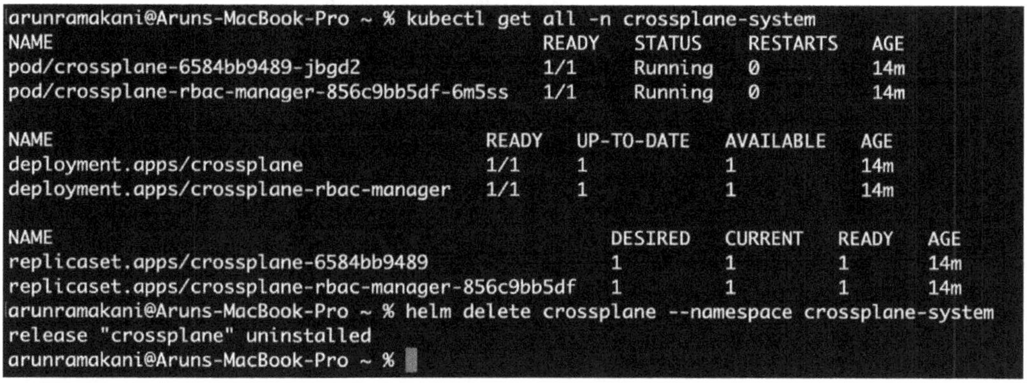

```
arunramakani@Aruns-MacBook-Pro ~ % kubectl get all -n crossplane-system
NAME                                            READY    STATUS     RESTARTS    AGE
pod/crossplane-6584bb9489-jbgd2                 1/1      Running    0           14m
pod/crossplane-rbac-manager-856c9bb5df-6m5ss    1/1      Running    0           14m

NAME                                      READY    UP-TO-DATE    AVAILABLE    AGE
deployment.apps/crossplane                1/1      1             1            14m
deployment.apps/crossplane-rbac-manager   1/1      1             1            14m

NAME                                                DESIRED    CURRENT    READY    AGE
replicaset.apps/crossplane-6584bb9489               1          1          1        14m
replicaset.apps/crossplane-rbac-manager-856c9bb5df  1          1          1        14m
arunramakani@Aruns-MacBook-Pro ~ % helm delete crossplane --namespace crossplane-system
release "crossplane" uninstalled
arunramakani@Aruns-MacBook-Pro ~ %
```

Figure 3.8 – Crossplane installation health

Now that we have installed Crossplane, we will learn how to install and configure the providers in the following section.

Installing and configuring providers

Once we have the Crossplane core component installed in the Kubernetes cluster, the next step is installing and configuring the Crossplane provider. We will install and configure the GCP provider, which is the scope of our tutorial. We can do this in a three-step process:

1. Setting up a cloud account
2. Installing a provider
3. Configuring the provider

We will look at each of these aspects in detail with a step-by-step guide in the following sections.

Setting up a cloud account

We need to have a Google Cloud account with the project and billing setup enabled. Google Cloud offers $300 credit for new users to learn and experiment for a maximum of 3 months, provided you have a credit card. It will be more than enough for us to learn Crossplane and other infrastructure automation concepts. All we need to do is log on to the Google Cloud account to fill out a form and start the free trial. The next step is to create a separate project space for us to experiment with Crossplane. You can create a new project by clicking the project dropdown on the top bar of the GCP console and clicking on **NEW PROJECT**, as shown in the following screenshot:

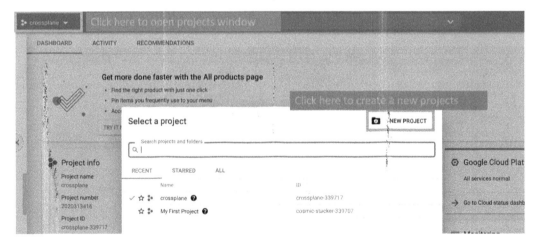

Figure 3.9 – Creating a new GCP project

Once we have a GCP account, free credits, and the project started, the next step is to enable all the needed cloud API access. We are going to do that with the gcloud CLI. Download the CLI from https://cloud.google.com/sdk/docs/install and follow these steps:

1. Install the gcloud CLI after the download:

    ```
    ./google-cloud-sdk/install.sh
    ```

2. Initialize the SDK with the cloud credentials and project:

    ```
    gcloud init
    ```

gcloud init will redirect us to a web browser to authenticate with the Google account. Once we can successfully provide necessary access, we must select the project when asked for it by the CLI. Now, we should be able to enable the required cloud services with the following commands:

```
# Enable Kubernetes APIs , CloudSQL APIs, Network APIs and
Compute APIs

gcloud services enable container.googleapis.com

gcloud services enable sqladmin.googleapis.com

gcloud services enable compute.googleapis.com

gcloud services enable servicenetworking.googleapis.com
```

A screenshot of the result is as follows:

```
arunramakani@Aruns-MacBook-Pro ~ % gcloud services enable container.googleapis.com
Operation "operations/acf.p2-955276575103-dbeff1e8-c0a3-434c-b42a-1005a8dd8c43" finished successfully.
arunramakani@Aruns-MacBook-Pro ~ % gcloud services enable sqladmin.googleapis.com
Operation "operations/acf.p2-955276575103-a1c04a94-6e38-44e3-b6a4-559f95e9612b" finished successfully.
arunramakani@Aruns-MacBook-Pro ~ % gcloud services enable compute.googleapis.com
arunramakani@Aruns-MacBook-Pro ~ % gcloud services enable servicenetworking.googleapis.com
Operation "operations/acf.p2-955276575103-d30d32f0-0e45-4d10-afec-4d41b1f50b7a" finished successfully.
arunramakani@Aruns-MacBook-Pro ~ %
```

Figure 3.10 – Enabling the cloud APIs

Once we are done enabling the API access for the project, the next step is to create a service account and extract the credentials required to set up the GCP Crossplane provider. It involves a few steps:

1. Find the project ID to use in the necessary upcoming commands:

   ```
   gcloud projects list --format='value(project_id)'
   ```

2. Create the service account, get the name, and enable the required roles on the service account. Note that `crossplane-330620` in all the following commands is my Google project ID. You will replace that with your Google project ID. Use the preceding command to explore the list of your project IDs. Similarly, `crossplane-service-account@crossplane-330620.iam.gserviceaccount.com` is the name of the created service account in my GCP environment. Replace this with your service account name. You can get your service account name by executing the second command in the following code block to list the service accounts:

   ```
   # Create service account
   gcloud iam service-accounts create crossplane-service-
   account --display-name "crossplane service account"
   --project=crossplane-330620


   # Get the name of the service account
   gcloud iam service-accounts list --filter="email ~
   ^crossplane-service-account" --format='value(email)'


   # Add required IAM role to the service account
   gcloud projects add-iam-policy-binding crossplane-330620
   --member "serviceAccount:crossplane-service-account@
   crossplane-330620.iam.gserviceaccount.com" --role="roles/
   iam.serviceAccountUser"


   gcloud projects add-iam-policy-binding crossplane-330620
   --member "serviceAccount:crossplane-service-account@
   crossplane-330620.iam.gserviceaccount.com" --role="roles/
   cloudsql.admin"


   gcloud projects add-iam-policy-binding crossplane-330620
   --member "serviceAccount:crossplane-service-account@
   ```

```
crossplane-330620.iam.gserviceaccount.com" --role="roles/
container.admin"

gcloud projects add-iam-policy-binding crossplane-330620
--member "serviceAccount:crossplane-service-account@
crossplane-330620.iam.gserviceaccount.com" --role="roles/
redis.admin"

gcloud projects add-iam-policy-binding crossplane-330620
--member "serviceAccount:crossplane-service-account@
crossplane-330620.iam.gserviceaccount.com" --role="roles/
compute.networkAdmin"

gcloud projects add-iam-policy-binding crossplane-330620
--member "serviceAccount:crossplane-service-account@
crossplane-330620.iam.gserviceaccount.com" --role="roles/
storage.admin"
```

3. To extract the service account file, use the first command, as follows:

```
gcloud iam service-accounts keys create crossplane-
service-account.json --iam-account crossplane-service-
account@crossplane-330620.iam.gserviceaccount.com
```

4. Finally, create a Kubernetes Secret with the service account file. The Secret will be referred to when we make a provider configuration, which you will see in the next section:

```
kubectl create secret generic gcp-account -n crossplane-
system --from-file=creds=./crossplane-service-account.
json
```

Installing a provider

Let's now install the GCP Crossplane provider. We have to run the following provider resource YAML with the latest GCP provider container image version. The current available stable version is v0.18.0. Update the version accordingly when you are executing:

```
apiVersion: pkg.crossplane.io/v1
kind: Provider
metadata:
```

```
  name: provider-gcp
spec:
  package: crossplane/provider-gcp:v0.18.0
```

Apply the YAML and validate whether the provider container is running:

```
kubectl apply -f Provider.yaml
```

```
kubectl get all -n crossplane-system
```

A screenshot of the result is as follows:

```
arunramakani@Aruns-MacBook-Pro Downloads % kubectl get all -n crossplane-system
NAME                                            READY   STATUS    RESTARTS   AGE
pod/crossplane-6584bb9489-vf7ln                 1/1     Running   1          15h
pod/crossplane-rbac-manager-856c9bb5df-kwtt9    1/1     Running   0          15h
pod/provider-gcp-3d69b3bf2649-578c98cf5f-w4565  1/1     Running   0          3m5s

NAME                                            READY   UP-TO-DATE   AVAILABLE   AGE
deployment.apps/crossplane                      1/1     1            1           15h
deployment.apps/crossplane-rbac-manager         1/1     1            1           15h
deployment.apps/provider-gcp-3d69b3bf2649       1/1     1            1           3m5s

NAME                                            DESIRED   CURRENT   READY   AGE
replicaset.apps/crossplane-6584bb9489           1         1         1       15h
replicaset.apps/crossplane-rbac-manager-856c9bb5df  1     1         1       15h
replicaset.apps/provider-gcp-3d69b3bf2649-578c98cf5f  1   1         1       3m5s
```

Figure 3.11 – Running the GCP provider

Configuring the provider

The final step is to set up provider configuration for authentication using the `ProviderConfig` Crossplane API. Preparing the credentials can be slightly different for each provider, depending on the construct for authentication by the infrastructure vendor. In the case of GCP, it will be a service account, it's a service principle for Azure, an IAM for AWS, and so on. The following YAML will configure credentials for the GCP provider:

```
apiVersion: gcp.crossplane.io/v1beta1
kind: ProviderConfig
metadata:
  name: gcp-credentials-project-1
spec:
```

```
projectID: crossplane-330620
credentials:
  source: Secret
  secretRef:
    namespace: crossplane-system
    name: gcp-account
    key: service-account
```

A couple of crucial aspects of GCP provider configuration are the project ID reference and the service account credentials. We need to provide our GCP project ID under projectID:. We will not have this configuration element for other infrastructure provider configurations such as AWS. Note that the provider configuration mentioned previously references the Kubernetes Secrets. Apply the YAML and validate the provider configuration with the following commands:

```
kubectl apply -f providerconfig.yaml
```

```
kubectl get providerconfig
```

A screenshot of the result is as follows:

```
arunramakani@Aruns-MacBook-Pro Downloads % kubectl apply -f providerconfig.yaml
providerconfig.gcp.crossplane.io/gcp-credentials-project-1 created
arunramakani@Aruns-MacBook-Pro Downloads %
arunramakani@Aruns-MacBook-Pro Downloads %
arunramakani@Aruns-MacBook-Pro Downloads % kubectl get providerconfig
NAME                       PROJECT-ID          AGE
gcp-credentials-project-1  crossplane-330620   5m43s
```

Figure 3.12 – The GCP provider config

With this, we are concluding Crossplane installation, GCP provider setup, and configuration. Other provider setups also mostly look like GCP. Now, the environment is ready for provisioning GCP resources with an MR, or we can start composing our XR API above the MR.

Multiple provider configuration

We can have multiple ProviderConfig configured against a provider. It's like having many credentials or cloud accounts to access the cloud platform and choosing the appropriate credentials based on the given context. When provisioning the infrastructure resources with an MR or XR, we specify `providerConfigRef:` to determine which ProviderConfig to use in the given context. If `providerConfigRef:` is not available in an MR or XR, Crossplane refers to the ProviderConfig named `default`. This way of organizing infrastructure resources under different credentials can help us manage infrastructure billing and maintain resources concerning the organizational structure in groups. The following YAML will provision POSTGRES from GCP using the provider config named `gcp-credentials-project-1`, which we created in the preceding section:

```
apiVersion: database.gcp.crossplane.io/v1beta1
kind: CloudSQLInstance
metadata:
  name: my-GCP-DB
spec:
  # Reference to use a specific provider config
  providerConfigRef:
    name: gcp-credentials-project-1
  forProvider:
    databaseVersion: POSTGRES_9_6
    region: asia-south2
    settings:
      tier: db-n1-standard-1
      dataDiskSizeGb: 10
```

The following figure represents multiple teams using different provider configs:

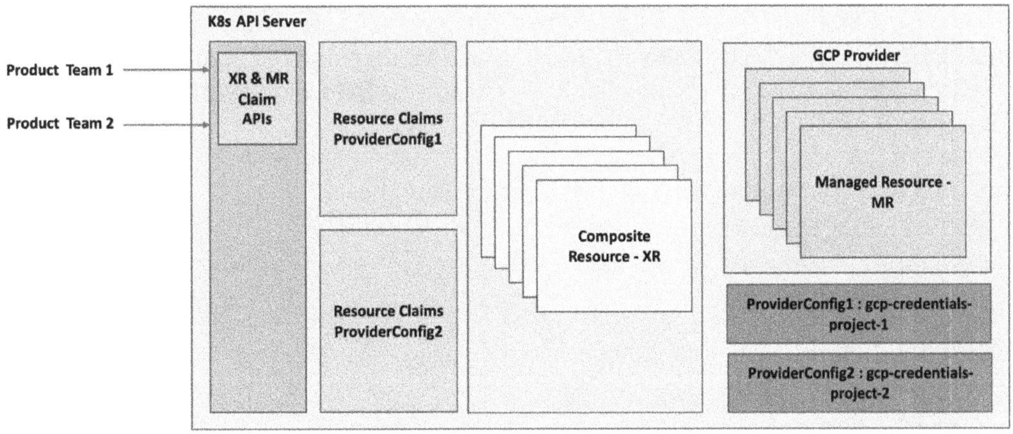

Figure 3.13 – Multiple provider configuration

An example of POSTGRES provisioning

It's time that we go through an actual provisioning experience. We will create
a POSTGRES instance in GCP using CloudSQLInstance, an MR available in the GCP
Crossplane provider. Direct provisioning of infrastructure resources with an MR or XR
is not a good strategy. Instead, we should be using the claim object created with XR for
provisioning. The next chapter is devoted to covering these topics. Currently, we are using
an MR for resource provisioning to understand the basic building blocks of Crossplane.
We are going to use the following attributes while provisioning the resource:

- The name of the resource provisioned should be db-gcp.

- The region of provisioning will be us-central.

- We will request POSTGRES version 9.6 – POSTGRES_9_6.

- The size of the data disk should be 20 GB.

- The GCP tier of the database should be db-g1-small.

- Create the resource under the gcp-credentials-project-1
  provider configuration.

- Database credentials should go to a crossplane-system namespace with a
  Secret named db-conn.

These are just a few possible parameters. The complete API documentation for CloudSQLInstance is available at `https://doc.crds.dev/github.com/crossplane/provider-gcp/database.gcp.crossplane.io/CloudSQLInstance/v1beta1@v0.18.0`. Note that there is a required tag against a few API parameters, which are mandatory in YAML. The following YAML code will provision GCP POSTGRES with the mentioned attributes:

```yaml
apiVersion: database.gcp.crossplane.io/v1beta1
kind: CloudSQLInstance
metadata:
  name: db-gcp
spec:
  providerConfigRef:
    name: gcp-credentials-project-1
  writeConnectionSecretToRef:
    namespace: crossplane-system
    name: db-conn
  forProvider:
    databaseVersion: POSTGRES_9_6
    region: us-central
    settings:
      tier: db-g1-small
      dataDiskSizeGb: 20
```

Once you apply the YAML code, you can see the resources provisioned in the GCP console. Use the following command to look at the states. Note that, initially, we will see the status as pending creation (`PENDING_CREATE`), and it will soon become runnable (`RUNNABLE`). Also, we can see that the database credentials are available in the Secrets:

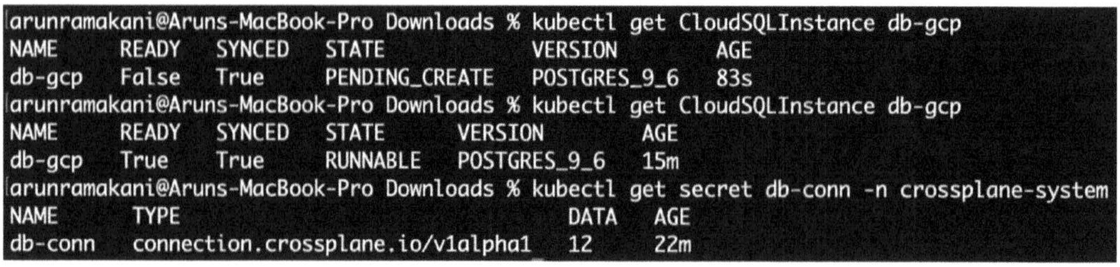

```
arunramakani@Aruns-MacBook-Pro Downloads % kubectl get CloudSQLInstance db-gcp
NAME      READY    SYNCED    STATE            VERSION        AGE
db-gcp    False    True      PENDING_CREATE   POSTGRES_9_6   83s
arunramakani@Aruns-MacBook-Pro Downloads % kubectl get CloudSQLInstance db-gcp
NAME      READY    SYNCED    STATE        VERSION        AGE
db-gcp    True     True      RUNNABLE     POSTGRES_9_6   15m
arunramakani@Aruns-MacBook-Pro Downloads % kubectl get secret db-conn -n crossplane-system
NAME       TYPE                                DATA    AGE
db-conn    connection.crossplane.io/v1alpha1   12      22m
```

Figure 3.14 – Database provisioning

Suppose we look at the database details in the GCP console and change the database machine type. Crossplane will restore it to the original tier mentioned in the YAML code. It was fun to try that and see the status change. Initially, it will go into maintenance mode when we change the console, and then Crossplane will realize that something was changed to put the resource out of sync. Then, Crossplane will bring it back to its original state. Please refer to the following screenshot and go through values in each column (STATE, SYNCED, and READY):

```
arunramakani@Aruns-MacBook-Pro Downloads % kubectl get CloudSQLInstance db-gcp
NAME      READY   SYNCED   STATE         VERSION        AGE
db-gcp    False   False    MAINTENANCE   POSTGRES_9_6   40m
arunramakani@Aruns-MacBook-Pro Downloads % kubectl get CloudSQLInstance db-gcp
NAME      READY   SYNCED   STATE         VERSION        AGE
db-gcp    True    False    RUNNABLE      POSTGRES_9_6   41m
arunramakani@Aruns-MacBook-Pro Downloads % kubectl get CloudSQLInstance db-gcp
NAME      READY   SYNCED   STATE         VERSION        AGE
db-gcp    False   False    MAINTENANCE   POSTGRES_9_6   44m
arunramakani@Aruns-MacBook-Pro Downloads % kubectl get CloudSQLInstance db-gcp
NAME      READY   SYNCED   STATE         VERSION        AGE
db-gcp    True    False    RUNNABLE      POSTGRES_9_6   45m
arunramakani@Aruns-MacBook-Pro Downloads % kubectl get CloudSQLInstance db-gcp
NAME      READY   SYNCED   STATE         VERSION        AGE
db-gcp    True    False    RUNNABLE      POSTGRES_9_6   46m
```

Figure 3.15 – The reconciliation loop

And finally, the resources will sync. We can clean up the provisioned resources based on the deletion policy:

```
arunramakani@Aruns-MacBook-Pro Downloads % kubectl get CloudSQLInstance db-gcp
NAME      READY   SYNCED   STATE         VERSION        AGE
db-gcp    True    True     RUNNABLE      POSTGRES_9_6   51m
arunramakani@Aruns-MacBook-Pro Downloads % kubectl delete CloudSQLInstance db-gcp
cloudsqlinstance.database.gcp.crossplane.io "db-gcp" deleted
```

Figure 3.16 – The final state

> **Tip**
>
> If you want to create the resource in the vendor infrastructure with a name different from that of the Crossplane resource claim, use the crossplane.io/external-name:my-special-name annotation in the metadata section.

All the examples discussed throughout this book can be referred to at https://github.com/PacktPublishing/End-to-End-Automation-with-Kubernetes-and-Crossplane.

Summary

We started with understanding CRDs and custom controllers, and then the Kubernetes resource extension point concept, which are the building blocks of Crossplane. Then, we went through various Crossplane components, their architecture, and how those components fit together. Finally, we undertook some hands-on work by installing Crossplane and its GCP provider, and experimenting with Postgres database provisioning. Now we know how Crossplane works end to end, which brings us to the end of the chapter. In the next chapter, we will learn advanced Crossplane concepts.

4

Composing Infrastructure with Crossplane

Composing is a powerful construct of Crossplane that makes it unique among its peers, such as the Open Service Broker API or AWS Controllers for Kubernetes. The ability to organize infrastructure recipes in a no-code way perfectly matches the organization's agile expectation of building a lean platform team. This chapter will take us on a journey to learn about composing from end to end. We will start with a detailed understanding of how Crossplane **Composite Resources** (**XRs**) work and then cover a hands-on journey to build an XR step by step.

The following are the topics covered in this chapter:

- Feeling like an API developer
- How do XRs work?
- Postprovisioning of an XR
- Preprovisioned resources
- Building an XR
- Troubleshooting

Feeling like an API developer

Traditionally, infrastructure engineers know the most profound infrastructure configuration options and different infrastructure setup patterns. But they may not have experience in building APIs. Building an infrastructure platform with Crossplane will be a shift from these usual ways. Modern infrastructure platform developers should have both pieces of knowledge, that is, infrastructure and API engineering. Building infrastructure APIs as a platform developer means implementing the following aspects:

- Evolving the APIs as time passes by. This involves introducing new APIs, updating an existing API version, and deprecating the old APIs.

- Applying API cross-cutting concerns for consuming product teams, such as authentication, authorization, caching, and auditing.

- Encapsulating different infrastructure policies within the APIs.

- Building reusable infrastructure recipes used across teams. For example, some product teams might develop their applications with the **MEAN** stack (**MongoDB, Express.js, AngularJS, and Node.js**). We might be interested in developing infrastructure provisioning for this stack as a template API.

- Building the required shared infrastructure used across teams. For example, we might want to provision a virtual private network shared by different infrastructure resources.

- Achieving and evolving correct API boundaries considering the different infrastructure recipes and shared infrastructure. We must perform trade-off analysis to deal with conflicting concerns between infrastructure recipes and shared infrastructure.

Infrastructure recipes and shared infrastructure are vital elements in API-bounded context trade-offs. We will examine this in detail in an upcoming chapter. The following figure represents the nuances of API infrastructure engineering:

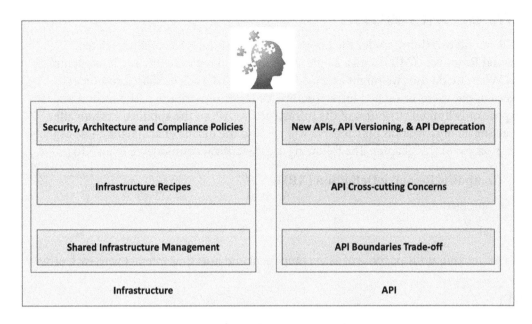

Figure 4.1 – API infrastructure engineering

We are looking at these aspects to understand XR architecture in the best possible way. Every element of the Crossplane composite is designed to cover infrastructure engineering practices from the perspective of an API.

> **Tip**
>
> We can use the learnings from microservices architecture pattern to define infrastructure API boundaries. There is no perfect boundary, and every design option will have advantages and disadvantages. In *Chapter 6, More Crossplane Patterns*, we can look for ways to adopt microservices with the Crossplane infrastructure platform.

How do XRs work?

An XR can do two things under the hood. The first purpose is to combine related **Managed Resources** (**MRs**) into a single stack and build reusable infrastructure template APIs. When we do this, we might apply different patterns, such as shared resources between applications or cached infrastructure for faster provisioning. The second one is to expose only limited attributes of the infrastructure API to the application team after abstracting all organization policies. We will get into the details of achieving these aspects as we progress in this chapter. The following are the critical components in an XR:

- **Composite Resource Definition** (**XRD**)
- Composition
- Claim

Let's start looking at the purpose of each component and how they interact with each other.

XRD

The **XRD** is the schema defining the infrastructure API specification. It is best to describe an XRD first. Fixing the API specification first will force us to think about end users' needs and the different ways they will consume the API. We will also apply all the organization policies to decide what fields are to be exposed to an application team. It will clearly set the scope and the boundary of the API. `CompositeResourceDefinition` is the Crossplane configuration element used to define an XRD. Creating this configuration is like writing an OpenAPI Specification or Swagger API definition. The following are the critical aspects of the `CompositeResourceDefinition` configuration YAML:

- **The XR names**: These will be the first to define, representing the infrastructure API name in singular and plural format. The singular name will eventually become the `kind` attribute of the new API. Note that it's standard practice to use *X* as a prefix for the XR name.

- **API group**: This will help us to group the API logically, avoid naming conflicts, and manage authorization.

- **Metadata name**: Metadata is a string value constructed in a standard format. It is the concatenation of the plural name (plural name of the XR) and a dot followed by the API group (the group under which we what to classify the XR resource). In other words, the string follows this template: `<resource plural name>.<API group>`.

- **versions**: This is an XRD configuration construct that will help us to manage the API versions. The `versions` element is an array and can hold configuration for multiple versions of the same XR API. Typically, when we start, we will have just one version. As time progresses, we will increment the API version with changes. The old version can become a technical debt to deprecate later.

- **served and referenceable**: These are a couple of mandatory Boolean attributes for every defined version. The `served` element will indicate whether the XR API is served with the given version. The `referenceable` flag will determine whether we can define an implementation for the given API version. We can look at version management and these attributes in more depth in *Chapter 5, Exploring Infrastructure Platform Patterns*. For now, both flags will be `true` when we have only one version defined in the XRD.

- **Schema**: This is a section under each version covering the actual OpenAPI specification. It covers details such as parameter lists, data types, and required parameters.

- **Connection secret keys**: This will hold the list of keys that need to be created and populated in the Kubernetes Secrets after the resource provisioning.

- **Composition reference**: These parameters influence which resource-composing implementation is to be used on specific infrastructure API calls. In other words, we could have multiple API implementations for the given XRD, and this section of the XRD configuration will help to define the default implementation or enforced implementation. `DefaultCompositionRef` and `EnforcedCompositionRef` are a couple of attributes providing this flexibility.

- **Claim names**: These are optional parameters that create a proxy API for the given XR API with the specified name. Applying the claim object's create, delete, and update action will create, delete, and update the underlying XR. Claims are a critical component in Crossplane, and we will look at that in a dedicated topic shortly in this chapter.

The XRD is nothing but an opinionated **Custom Resource Definition** (**CRD**), and many parts of the configuration look like a CRD. These are just a few possible parameters. We will look at a few more parameters as we progress through the book. The complete API documentation is available at `https://doc.crds.dev/github.com/crossplane/crossplane`.

> **Tip**
> We are looking at v1.5.1 of the Crossplane documentation, which is the latest at the time of writing this chapter. Refer to the latest version at the time of reading for more accurate details.

Note that some of the configurations discussed previously are not part of the following YAML, such as `DefaultCompositionRef` and `ConnectionSecretKeys`. These configurations are injected by Crossplane with default behavior if not specified. Refer to the following YAML for an example:

```
apiVersion: apiextensions.crossplane.io/v1
kind: CompositeResourceDefinition
metadata:
  #'<plural>.<group>'
  name: xclouddbs.book.imarunrk.com
spec:
  # API group
  group: book.imarunrk.com
  # Singular name and plural name.
  names:
    kind: xclouddb
    plural: xclouddbs
  # Optional parameter to create namespace proxy claim API
  claimNames:
    kind: Clouddb
    plural: Clouddbs
  # Start from alpha to beta to production to deprecated.
  versions:
  - name: v1
    # Is the specific version actively served
    served: true
    # Can the version be referenced from an API implementation
    referenceable: true
    # OpenAPI schema
    schema:
      openAPIV3Schema:
        type: object
        properties:
          spec:
            type: object
```

```
        properties:
          parameters:
            type: object
            properties:
              storageSize:
                type: integer
            required:
            - storageSize
        required:
        - parameters
```

Once we are done with the API specification, the next step is to build the API implementation. Composition is the Crossplane construct used for providing API implementation.

Composition

The composition will link one or more MRs with an XR API. When we create, update, and delete an XR, the same operation will happen on all the linked MRs. We can consider XRD as the CRD and composition as the custom controller implementation. The following diagram represents how XR, XRD, composition, and MRs are related:

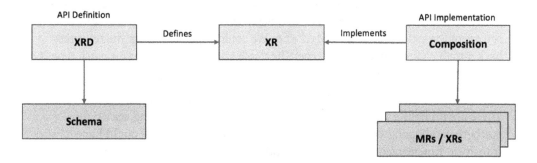

Figure 4.2 – XRM, composition, and XR

Tip

We have referred to XR in this book in two contexts. We can use XR to refer to a new infrastructure API that we are building. Also, the composition resources list can hold both an MR and an existing XR. We will also refer to an XR from that context. Look at *Figure 4.2* where XR is referred to in both dimensions.

Let's look at some of the crucial elements from the composition configuration:

- **CompositeTypeRef**: This is an attribute that will help us to map a specific XR version with the current composition. `kind` and `apiVersion` are the two configuration elements defined under `CompositeTypeRef`. While `kind` specifies the XR name, `apiVersion` will refer to a specific version defined in the XRD. The mapped version should be configured as referenceable in the XRD.

- **WriteConnectionSecretsToNamespace**: This will specify the namespace for storing the connection Secrets.

- **Resources**: This section is an array that holds the list of MRs to be created, updated, and deleted when someone creates, updates, and deletes the XR. We can even define another XR under this section. It is a mandatory section, and we should define at least one resource, either an MR or XR. The base is the critical object under each resource that holds the XR/MR configuration template.

- **Patches**: This section under a given resource will be helpful to overlay the API input attributes to the composing resource (MR/XR) attributes. This section is optional and an array where we can specify multiple patches. There are many predefined types of patches. `FromCompositeFieldPath` is the default type and is used most frequently. It is helpful to patch an attribute from the XR into the composition resource base template, that is, feeding the user input into the composing resources. `FromFieldPath` and `ToFieldPath` are the subattributes that perform the actual patching. There is a patch type called `ToCompositeFieldPath`, which does the reverse of `FromCompositeFieldPath`. We could copy fields from the resources back into the XR using this patch type. The `CombineFromComposite` patch type is the most suitable option when combining multiple attributes.

- **Transforms**: These are optional elements helpful in computing the patched fields. These are predefined functions, such as `convert` for typecasting, `math` for mathematical operations, and `map` for key-value operations. We could have a list of transform functions on a given patch, and they are executed in the order specified in the configuration. Both patches and transforms are vital patterns. We will look at different configuration examples for patches and transforms throughout the book.

- **Policy**: These are under each patch and will determine the patching behavior. We can mandate the patch path presence because the default behavior is to skip the patch if the field is absent. Also, we can configure the behavior of merge when the patching is performed over an object.

- **ConnectionDetails**: These are specified under each resource and will hold the list of secret keys we want to propagate back into the XR.

- **ReadinessChecks**: These will allow us to define any custom readiness logic. If this section is not provided, the default behavior is to make the XR state ready when all the composing resources are ready.

- **PatchSets**: This is the final attribute that we will cover. Patch sets allow us to define a set of reusable patch functions that can be used across multiple resources. It's like a shared reusable function.

> **Tip**
>
> We have a few field path attributes when defining a composition. Values for these fields will follow standard JavaScript syntax to access JSON, for example, `spec.parameters.storageSize` or `spec.versions[0].name`.

We covered most of the configuration options available with the composition. Have a look at the Crossplane documentation for the complete list. The following figure represents the composition configuration options and the relationship between them:

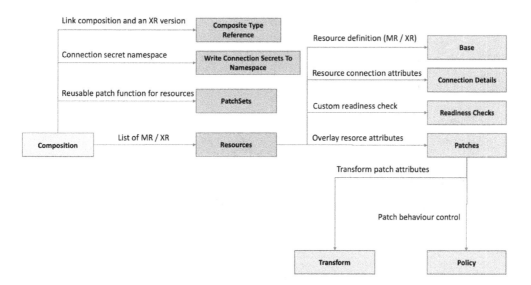

Figure 4.3 – Composition configuration

The following is a sample composition configuration YAML:

```
apiVersion: apiextensions.crossplane.io/v1
kind: Composition
metadata:
  name: xclouddb-composition
```

```yaml
spec:
  # Link Composition to a specific XR and version
  compositeTypeRef:
    apiVersion: xclouddb.book.imarunrk.com/v1
    kind: Xclouddb
  # Connection secrets namespace
  writeConnectionSecretsToNamespace: crossplane-system
  # List of composed MRs or XRs.
  resources:
  - name: clouddbInstance
    # Resource base template
    base:
      apiVersion: database.gcp.crossplane.io/v1beta1
      kind: CloudSQLInstance
      spec:
        forProvider:
          databaseVersion: POSTGRES_9_6
          region: us-central
          settings:
            tier: db-g1-small
            dataDiskSizeGb: 20
    # Resource patches
    patches:
    - type: FromCompositeFieldPath
      fromFieldPath: spec.parameters.storageSize
      toFieldPath: spec.forProvider.settings.dataDiskSizeGb
    # Resource secrets
    connectionDetails:
    - name: hostname
      fromConnectionSecretKey: hostname
```

We will cover an example with more configuration elements in the *Building an XR* section. An XRD version can have more than one composition, that is, one-to-many relationships between the XRD version and composition. It provides polymorphic behavior for our infrastructure API to work based on the context. For example, we could have different compositions defined for production and staging. The `CompositionRef` attribute defined in the XR can refer to a specific composition. Instead of `CompositionRef`, we can also use `CompositionSelector` to match the compositions based on labels.

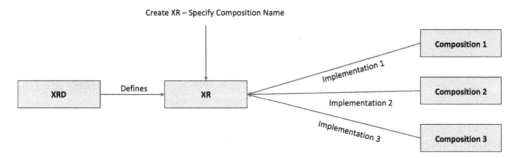

Figure 4.4 – XR and composition relation

In the next section, we will look at **XR claims**, also known as **claims**.

Claim

A claim is a proxy API to the XR, created by providing claim name attributes in XRD configurations. As a general practice, we provide the exact name of the XR after removing the initial X. In the preceding example, `xclouddb` is the XR name and `Clouddb` is the claim name but following such naming conventions is not mandatory. Claims are very similar to the XR, and it might tempt us to think that it's an unnecessary proxy layer. Having a claim is helpful in many ways, such as the following:

- XRs are cluster-level resources, while the claims are namespace level. It enables us to create namespace-level authorization. For example, we can assign different permissions for different product teams based on their namespace ownership.

- We can keep some of the XR only as a private API at the cluster level for the platform team's use. For example, the platform team may not be interested in exposing the XR API that creates a virtual private network.

- It's not ideal to manage some of the resources at the namespace level as they are shared between teams and do not fit into the context.

- We can also use this pattern to support the preprovisioning of infrastructure. A claim can just reference itself with a preprovisioned XR infrastructure, keeping the provisioning time low. It is very similar to caching.

The following figure represents how claims, XR, XRD, composition, and MRs are related, giving an end-to-end view of how the whole concept works:

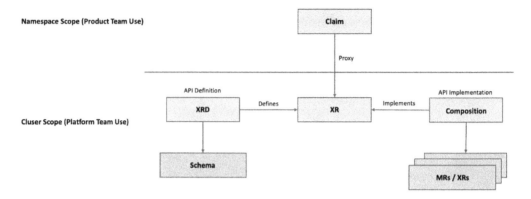

Figure 4.5 – How does composition work?

The following are the sample claim and XR YAML. The claim YAML is as follows:

```
apiVersion: book.imarunrk.com/v1
# Kind name matches the singular claim name in the XRD
kind: Clouddb
metadata:
  name: cloud-db
spec:
  # Parameters to be mapped and patched in the composition
  parameters:
    storageSize: 20
  # Name of the composition to be used
  compositionRef:
    name: xclouddb-composition
  writeConnectionSecretToRef:
    namespace: crossplane-system
    name: db-conn
```

A namespace is not part of the preceding claim YAML. Hence, it will create the resource in the default namespace, the Kubernetes standard. An equivalent XR YAML to the preceding claim YAML is as follows:

```
apiVersion: book.imarunrk.com/v1
kind: XClouddb
```

```
metadata:
  name: cloud-db
spec:
  parameters:
    storageSize: 20
  compositionRef:
    name: xclouddb-composition
  writeConnectionSecretToRef:
    namespace: crossplane-system
    name: db-conn
```

Note that the XR is always created at the cluster level and namespace configuration under metadata is not applicable. We can look at a more detailed claim and XR configurations in the *Building an XR* section. Let's explore a few more XR, XRD, composition, and claim configurations from the perspective of postprovisioning requirements.

Postprovisioning of an XR

After performing CRUD operations over a claim or XR resource, the following are some critical aspects to bring the API request to a close:

- Readiness check

- Patch status

- Propagating the credentials back

Let's start with learning about readiness checks.

Readiness check

The XR state will be ready by default when all the underlying resources are ready. Every resource element in the composition can define its custom readiness logic. Let's look at a few of the custom readiness check configurations. If you want to match one of the composing resource status fields to a predefined string, use MatchString. A sample configuration for MatchString is as follows:

```
- type: MatchString
  fieldPath: status.atProvider.state
  matchString: "Online"
```

`MatchInteger` will perform a similar function when two integers are matched. The following sample configuration will check the `state` attribute with integer `1`:

```
- type: MatchInteger
  fieldPath: status.atProvider.state
  matchInteger: 1
```

Use the `None` type to consider the readiness as soon as the resource is available:

```
- type: None
```

Use `NonEmpty` to make the resource ready as soon as some value exists in the field of our choice. The following example will make the readiness true as soon as some value exists under the mentioned field path:

```
- type: NonEmpty
  fieldPath: status.atProvider.state
```

In the next section, we will look at an example of patching a status attribute after resource provisioning. Note that `fieldPath` falls under the `status` attribute. These are the attributes filled by MR during resource provisioning based on the values it gets back from the cloud provider.

Patch status

`ToCompositeFieldPath` is a patch type for copying any attribute from a specific composed resource back into the XR. Generally, we use it to copy the status fields. We can look at these as a way to define the API response. While there is a set of existing default status fields, patched fields are custom defined to enhance our debugging, monitoring, and audit activities. First, we need to define the state fields as a part of openAPIV3Schema in the XRD to make the new status fields available in the XR. The next step is to define a patch under the specific composing resource. The following patch will copy the current disk size of the CloudSQLInstance to the XR:

```
- type: ToCompositeFieldPath
  fromFieldPath: status.atProvider.currentDiskSize
  toFieldPath: status.dbDiskSize
```

We can also use the `CombineToComposite` patch type if we need to copy a combination of multiple fields.

Propagating credentials back

We can see that the connection secret-related configuration is part of the XRD, XR, claim, and composition. We must understand the relationship between these configurations to configure it correctly and get it working:

- Define the list of connection secret keys in the XRD using the `ConnectionSecretKeys` configuration.

- Configure the composing resources to define how to populate connection keys defined in the XRD. Connection details configuration can be of different types. The `FromConnectionSecretKey` type is correct when copying the secret from an existing secret key. We have the `FromFieldPath` type for copying the connection details from one of the composing resource fields.

- The claim or XR should save the Secrets using the `WriteConnectionSecretToRef` configuration.

The following diagram can help create a mind map of these configurations:

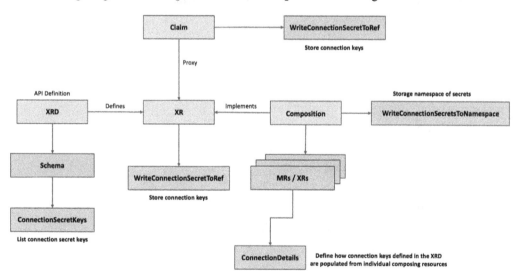

Figure 4.6 – Propagating the Secrets

The section covered different patterns that we can use with composition after the resources are provisioned. It is like customizing the API responses. Now we can look at the usefulness of reusing existing resources.

Preprovisioned resources

There are a few use cases where we may not create a new external resource and instead will reuse an existing provisioned resource. We will look at two such use cases in this section. The first use case is when we decide to cache the composed recourses because new resource provisioning may take too long to complete. The platform team can provision an XR and keep the resources in the resource pool. Then, the product team can claim these resources by adding the `ResourceRef` configuration under the spec of a claim YAML. With this pattern, we should ensure that the new claim attributes match the attributes in the existing pre-provisioned XR. If some of the attributes are different, Crossplane will try to update the XR specifications to match what is mentioned in the claim.

The second use case is about importing the existing resources from the external provider into the Crossplane. The `crossplane.io/external-name` annotation can help with this. Crossplane will look for an existing resource with the name mentioned in this configuration. The external name configuration mentioned in a claim will automatically be propagated into the XR. Still, it's our responsibility to patch this configuration into the composing resource. The following is a sample MR YAML where we onboard an existing VPC with the name `alpha-beta-vpc`:

```yaml
apiVersion: compute.gcp.crossplane.io/v1beta1
kind: Network
metadata:
  name: alpha-beta-vpc-crossplane-ref
  annotations:
    # Annotation to provide existing resource named
    crossplane.io/external-name: alpha-beta-vpc
spec:
  providerConfigRef:
    name: gcp-credentials-project-1
  # Provide the required parameters same as external resource.
  forProvider:
    autoCreateSubnetworks: true
```

Once you apply the YAML, you will see that it's ready for use in Crossplane. This can be seen in the following screenshot:

Figure 4.7 – VPC reference status

Note that the `alpha-beta-vpc` VPC is an existing VPC we created manually in GCP. What we achieve here is to map the manual resource to a Claim. The section covered different ways we can use preprovisioned resources with an XR/claim. The following section will be a hands-on journey to build an XR from scratch.

Building an XR

It's time to go through a hands-on journey to build an XR from scratch. We will start with writing down the infrastructure API requirement at a high level, then provide an API specification with XRD and finally provide an implementation with a composition. We will cover the API requirement in such a way as to learn most of the configuration discussed in this chapter.

The infrastructure API requirement

We will develop an API to provision a database from Google Cloud. The following are the compliance, architecture, and product team's requirements:

- **Compliance policy**: The provisioning should be done in the `us-central` region to comply with the data storage regulations from the government.

- **Architecture policy**: We should have two tiers of the database. For small, the disk size should be 20 GB, and it should be 40 GB for big.

- **Architecture policy**: The small tier's virtual machine should be `db-g1-small`, and `db-n1-standard-1` for the big tier.

- **Product team**: We should have the option to choose between Postgres and MySQL.

- **Product team**: We should specify the size in the XR with two enums (`SMALL` or `BIG`).

- **Platform team**: Patch the zone in which the database is created back into the XR/claim status field for monitoring requirements.

The next step is to write the XRD configuration YAML.

Creating the XRD

When defining the API specification with an XRD, the following configurations should be encoded into the YAML:

- Use `alpha-beta.imarunrk.com` as the API group to organize all APIs for alpha and beta teams.

- We will provide the XR name as `XGCPdb` and the claim name as `GCPdb`.

- We will start with a new API version, v1.

- Create `size` as an input parameter and `zone` as the response status attribute.

As the example XRD is oversized, we will cover only the schema definition here. Refer to the entire XRD file at `https://github.com/PacktPublishing/End-to-End-Automation-with-Kubernetes-and-Crossplane/blob/main/Chapter04/Hand-on-examples/Build-an-XR/xrd.yaml`. Without wasting much time, let's look at the schema:

```
schema:
  openAPIV3Schema:
    type: object
    properties:
      # Spec - defines the API input
      spec:
        type: object
        properties:
          parameters:
            type: object
            properties:
              # Size will be a user input
              size:
                type: string
            required:
            - size
        required:
        - parameters
      # status - the additional API output parameter
      status:
        type: object
        # Recourse zone - status patch parameter.
        properties:
          zone:
            description: DB zone.
            type: string
```

Save the YAML from GitHub and apply it to the cluster with `kubectl apply -f xrd.yaml`. Refer to the following screenshot, which shows successful XRD creation:

```
arunramakani@Aruns-MacBook-Pro gcpdb % kubectl apply -f xrd.yaml
compositeresourcedefinition.apiextensions.crossplane.io/xgcpdbs.alpha-beta.imarunrk.com created
arunramakani@Aruns-MacBook-Pro gcpdb % kubectl get xrd
NAME                               ESTABLISHED   OFFERED   AGE
xgcpdbs.alpha-beta.imarunrk.com    True          True      4s
```

Figure 4.8 – XRD creation

Note that the `ESTABLISHED` and `OFFERED` flags in the screenshot are `True`. This means that the XRD is created correctly. If these statuses are not `True`, use kubectl to describe the details of the XRD and look for an error.

Providing implementation

The next step is to provide an API implementation. As a part of the implementation, we should be providing a composition configuration. We will create two compositions, one for Postgres and the other for MySQL. It will be an example of the polymorphic behavior implementation. The following are the steps to remember when we build the composition YAML:

1. Refer to the v1 XRD API version with the `CompositeTypeRef` configuration.

2. Define the `CloudSQLInstance` configuration under the resource base.

3. Hardcode the region to `us-central1` to meet the compliance requirement.

4. The database tier and disk size will hold default values, but the patch configuration will overlay them using the `FromCompositeFieldPath` patch type.

5. Use the `Map` transformation to convert the `SMALL` tier size to the `db-g1-small` machine tier. Use the `Map` and `Convert` transformations to map the `SMALL` tier size to the 20 GB disk size.

6. Similar mapping will be done for the `BIG` configuration.

7. Patch the `GceZone` attribute from the MR status to the XR/claim for monitoring. We can achieve this using the `ToCompositeFieldPath` patch type.

8. Provide a mapping between the MR connection secret key to the XR/claim keys with the `ConnectionDetails` configuration.

We will look at the Postgres composition example in four parts. The XRD and resource definition section of the composition will look like the following configuration:

```
spec:
  # Refer to an XRD API version
  compositeTypeRef:
```

```
apiVersion: alpha-beta.imarunrk.com/v1
kind: XGCPdb
writeConnectionSecretsToNamespace: crossplane-system
resources:
  # Provide configuration for Postgres resource
- name: cloudsqlinstance
  base:
    apiVersion: database.gcp.crossplane.io/v1beta1
    kind: CloudSQLInstance
    spec:
      # reference to GCP credentials
      providerConfigRef:
        name: gcp-credentials-project-1
      forProvider:
        databaseVersion: POSTGRES_9_6
        # Compliance Policy
        region: us-central1
        settings:
          # These are default values
          # Architecture policies will be a patch
          tier: db-g1-small
          dataDiskSizeGb: 20
```

Read through the comments between the code snippets to understand concepts in detail.
The following configuration uses the map transform to patch the virtual machine tier:

```
- type: FromCompositeFieldPath
  fromFieldPath: spec.parameters.size
  toFieldPath: spec.forProvider.settings.tier
      # Use map transform
        # If the from-field value is BIG, then
        # the mapped to-field value is db-n1-standard-1
      transforms:
      - type: map
        map:
          BIG: db-n1-standard-1
          SMALL: db-g1-small
```

```
      policy:
          # return error if there is no field.
          fromFieldPath: Required
```

Next, we can look at the configuration to patch the disk size. The patch will have two transform operations. The first operation is to map the disk size, and the second one is to convert the mapped string to an integer:

```
- type: FromCompositeFieldPath
  fromFieldPath: spec.parameters.size
  toFieldPath: spec.forProvider.settings.dataDiskSizeGb
  # If the from-field value is BIG, then
  # the mapped to-field value is '40;
  # Apply the second transform to convert '40' to int
  transforms:
  - type: map
    map:
      BIG: "40"
      SMALL: "20"
  - type: convert
    convert:
      toType: int
  policy:
      # return error if there is no field.
      fromFieldPath: Required
```

Finally, the following patch adds the resource zone into the API response:

```
# Patch zone information back to the XR status
# No transformation or policy required
- type: ToCompositeFieldPath
  fromFieldPath: status.atProvider.gceZone
  toFieldPath: status.zone
```

The composition configuration for MySQL will be the same as the preceding configuration, excluding two changes. We should be changing the name of the composition in the metadata, and in the resource definition, we should change the database version to `MYSQL_5_7`. We can implement this with an additional parameter in the XR as well. Building two different compositions does not make sense when the difference is so small. We can capture the difference as a parameter in the XR. We are building two compositions, as an example. All composition examples and the upcoming claim examples are available for reference at `https://github.com/PacktPublishing/End-to-End-Automation-with-Kubernetes-and-Crossplane/tree/main/Chapter04/Hand-on-examples/Build-an-XR`.

Refer to the following screenshot, which shows the successful creation of both compositions:

```
arunramakani@Aruns-MacBook-Pro gcpdb % kubectl get composition
NAME        AGE
mysql       7s
postgres    41m
```

Figure 4.9 – Composition created

The final step is to use the claim API and create the database resources.

Provisioning the resources with a claim

Finally, we can start provisioning the GCP database with an XR or a claim. The `CompositionRef` configuration will specify which composition implementation to use. Note that the claims are namespace resources, and we provision them in the `alpha` namespace here. The following is a sample claim YAML for the MySQL database:

```
apiVersion: alpha-beta.imarunrk.com/v1
kind: GCPdb
metadata:
  # Claims in alpha namespace
  namespace: alpha
  name: mysql-db
spec:
  # Refer to the mysql composition
  compositionRef:
    name: mysql
  # save connection details as secret - db-conn2
  writeConnectionSecretToRef:
    name: db-conn2
```

```
    parameters:
      size: SMALL
```

The Postgres YAML as well will look similar with minor changes. Refer to the following screenshot, which shows a successful database creation:

```
arunramakani@Aruns-MacBook-Pro gcpdb % kubectl get cloudsqlinstance
NAME                         READY    SYNCED   . STATE       VERSION        AGE
mysql-db-lfqqw-tc4rh         True     True       RUNNABLE    MYSQL_5_7      5m47s
postgres-db-cvtnt-trcjg      True     True       RUNNABLE    POSTGRES_9_6   95m
arunramakani@Aruns-MacBook-Pro gcpdb % kubectl get gcpdb -n alpha
NAME           READY    CONNECTION-SECRET    AGE
mysql-db       True     db-conn2             5m50s
postgres-db    True     db-conn              95m
```

Figure 4.10 – Claim status

Note that the zone information is made available as the part of claim status:

```
Status:
    Conditions:
        Last Transition Time:    2021-12-05T23:48:01Z
        Reason:                  Available
        Status:                  True
        Type:                    Ready
    Connection Details:
        Last Published Time:     2021-12-05T23:48:01Z
    Zone:                        us-central1-f
```

Figure 4.11 – Zone information

This concludes the journey to build an XR. We will look at a few troubleshooting tips.

Troubleshooting

If we face issues with our infrastructure API, these tips could help us debug the problem in the best possible way:

- Status attributes and events are essential elements to debug issues. These details can be viewed by running `kubectl describe` on the given resource.

- When we start looking for issues, we take a top-down approach. This is because Crossplane follows the same convention as Kubernetes to hold the errors close to the resource where it happens.

- The debugging order will be *claim*, then *XR*, and then *each composing resource*. We should start with a claimed object. If we cannot locate the issue, we go deep into the XR and then the composing resources.

- `spec.resourceRef` from the claim description can help us to identify the XR name. Again, the same attribute can be used to find the composing resources from the XR.

Make an intentional mistake in the resource configuration of the composition to go through the debugging experience. You learn more when you debug issues. This concludes our troubleshooting section. Next, we will look at the chapter summary before moving on to the next chapter.

Summary

With this chapter, we covered one of the critical aspects of Crossplane, the XR. We started with understanding how an XR works and configuring an XR. Above all, we went through a hands-on journey to build a fresh infrastructure API from end to end. The chapter also covered some advanced XR configuration patterns and ways to approach debugging when there is an issue. This will be the base knowledge for what we will learn in the next chapter.

The next chapter will cover different advanced infrastructure platform patterns.

5
Exploring Infrastructure Platform Patterns

The success of running an infrastructure platform product with Crossplane depends on following a few principles and patterns as and when required. This chapter will explore some of these critical practices. We will also learn a few debugging skills while exploring the concepts. After learning the basics of the Crossplane in the last few chapters, this will be a place for learning advanced patterns that are key to building the state-of-the-art infrastructure platform for your organization. You will learn a few critical aspects of building robust XR APIs and debugging issues with ease.

The topics covered in this chapter are as follows:

- Evolving the APIs
- Nested and multi-resource XRs
- XRD detailed
- Managing external software resources

Evolving the APIs

Crossplane is primarily an API-based infrastructure automation platform. Changes to the APIs are inevitable as the business requirements and technology landscape evolve. We can classify these changes into three different buckets:

- API implementation change

- Non-breaking API contract change

- Breaking API contract change

Let's start with the API implementation change.

API implementation change

These changes are limited to the API implementation details without any changes to the contract. In other words, these are changes to Compositions YAML, a construct used by XR for API implementation. `CompositionRevision` is the Crossplane concept that will work with compositions to support such changes. If the `--enable-composition-revisions` flag is set while installing Crossplane, a `CompositionRevision` object is created with all the updates to composition. The name of the `CompositionRevision` object is autogenerated on every increment. The compositions are mutable objects that can change forever, but individual `CompositionRevision` is immutable. `Composition` and `CompositionRevision` are in one-to-many relationships. We will have only one `CompositionRevision` active at any given instance. The latest revision number will always be active, excluding the following scenario.

> Tip
> Each configuration state of the composition maps to a single `CompositionRevision`. Let's say we are in revision 2 and changing the composition configuration the same as the first revision. A new revision is not created. Instead, revision 1 becomes active, making revision 2 inactive.

In a Crossplane environment where the composition revision flag is enabled, we will have two attributes automatically added to every XR/Claim object by Crossplane. The following are attribute names and how they are used:

- `spec.compositionRevisionRef`: This will hold the name of `CompositionRevision` with which the resources are created.

- `spec.compositionUpdatePolicy`: This attribute will indicate whether the XR/Claim will automatically migrate to a new, available `CompositionRevision`. Manual and automatic are the two possible values, with automatic as the default value. If you would like to override the default behavior, add this attribute with a manual indicator in the XR/Claim configuration.

The following diagram represents how `Composition` and `CompositionRevision` work together to evolve infrastructure API implementation continuously:

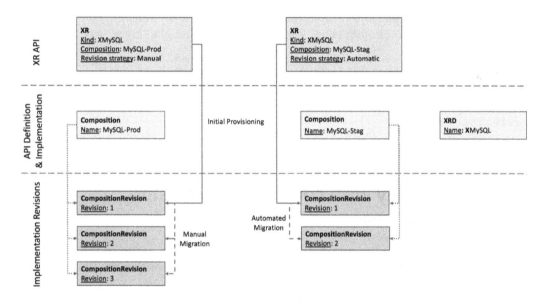

Figure 5.1 – Evolving compositions

To manually migrate the composition, update the `spec.compositionRevisionRef` configuration in the XR/Claim with the latest revision name. This specific design enables the separation of concerns between the platform API creator and consumers. Infrastructure API creators will update the compositions, and API consumers can choose their revision upgrade strategy. If you want a specific revision of composition to be used while creating an XR/Claim, explicitly mention the revision name under `spec.compositionRevisionRef`.

Let's look at some examples of such changes:

- **Bugfix**: Let's say we mapped an incorrect attribute to the XRD status field. The scope of the correct mapping is limited to changes in the respective Composition patch section.

- **Policy changes without contract change**: Adding a new infrastructure compliance policy to provide all new resources in a specific region.

- **Shared infrastructure**: Moving toward a shared **Virtual Private Cloud** (**VPC**) instead of dynamically provisioning a new VPC for all new XR/Claim requests.

The composition revision flag is not enabled by default. Use the `--enable-composition-revisions` argument with a Crossplane pod to enable composition revision. The following Helm command will set up/update the Crossplane environment with composition revision:

```
#Enable Composition revision in an existing environment
helm upgrade crossplane -namespace crossplane-system
crossplane-stable/crossplane -set args='{--enable-composition-
revisions}'
#Enable Composition revision in a new Crossplane setup
helm install crossplane -namespace crossplane-system
crossplane-stable/crossplane -set args='{--enable-composition-
revisions}'
```

The following section will look at composition revision with an example.

Hands-on journey with composition revision

Let's go through a hands-on journey to experience composition revision. The objectives of the exercise will be as follows:

- Building an XR API for GCP MySQL provisioning in a composition revision-enabled Crossplane environment

- Creating two MySQL instances with automated and manual composition revision policies

- Updating the Composition to change the calculation for database disk size

- Validating if the MySQL instance with automated revision policy automatically migrates to the latest composition revision

- Seeing that the MySQL instance with the manual revision policy does not migrate to the latest composition revision

- Finally, migrating the second MySQL instance manually to the latest composition revision

Let's use a simple XRD and composition to explore composition revision. The following is the XRD with just one parameter that takes the MySQL disk size:

```
apiVersion: apiextensions.crossplane.io/v1
kind: CompositeResourceDefinition
```

```
metadata:
  name: xmysqls.composition-revision.imarunrk.com
spec:
  group: composition-revision.imarunrk.com
  names:
    kind: XMySql
    plural: xmysqls
  claimNames:
    kind: MySql
    plural: mysqls
  versions:
  - name: v1
    served: true
    referenceable: true
    schema:
      openAPIV3Schema:
        type: object
        properties:
          spec:
            type: object
            properties:
              parameters:
                type: object
                properties:
                  size:
                    type: integer
                required:
                - size
            required:
            - parameters
```

The composition for the preceding XRD is as follows, which patches the `size` attribute from XR into the GCP CloudSQLInstance MR:

```
apiVersion: apiextensions.crossplane.io/v1
kind: Composition
metadata:
```

```
    name: gcp-mysql
spec:
  compositeTypeRef:
    apiVersion: composition-revision.imarunrk.com/v1
    kind: XMySql
  resources:
  - name: cloudsqlinstance
    base:
      apiVersion: database.gcp.crossplane.io/v1beta1
      kind: CloudSQLInstance
      spec:
        providerConfigRef:
          name: gcp-credentials-project-1
        forProvider:
          region: us-central1
          databaseVersion: MYSQL_5_7
          settings:
            tier: db-g1-small
            dataDiskSizeGb: 40
    patches:
    - type: FromCompositeFieldPath
      fromFieldPath: spec.parameters.size
      toFieldPath: spec.forProvider.settings.dataDiskSizeGb
```

Apply both the YAML to a target Crossplane cluster with composition revision enabled. You will see that `CompositionRevision` is created for the composition. Execute the following command to view all `CompositionRevision` for the given composition:

```
# List of revisions for Composition named gcp-mysql
kubectl get compositionrevision -l crossplane.io/composition-
name=gcp-mysql
```

Refer to the following screenshot with one revision object created for the gcp-mysql composition. Note that the `current` attribute is true for revision 1. It will change if we update the composition:

```
arunramakani@Aruns-MacBook-Pro Composition Revision % kubectl apply -f xrd.yaml
compositeresourcedefinition.apiextensions.crossplane.io/xmysqls.composition-revision.imarunrk.com created
arunramakani@Aruns-MacBook-Pro Composition Revision % kubectl apply -f Composition\ V1.yaml
composition.apiextensions.crossplane.io/gcp-mysql created
arunramakani@Aruns-MacBook-Pro Composition Revision % kubectl get compositionrevision -l crossplane.io/composition-name=gcp-mysql
NAME               REVISION   CURRENT   AGE
gcp-mysql-tlZp6    1          True      119s
```

Figure 5.2 – Composition Revision list

Now, let's provision two MySQL instances with the Claim API. An example of manual revision update policy configuration is as follows. The automated revision version of the YAML will be the same without the `compositionUpdatePolicy` parameter, which defaults to an automatic revision update:

```
apiVersion: composition-revision.imarunrk.com/v1
kind: MySql
metadata:
  namespace: alpha
  name: mysql-db-manual
spec:
  compositionUpdatePolicy: Manual
  compositionRef:
    name: gcp-mysql
  parameters:
    size: 10
```

You can refer to the following screenshot with two MySQL instances onboarded:

Figure 5.3 – MySQL claims

Now, update the composition patch with a transform function to multiply the disk size by four before patching. The patches section of the updated composition will look like the following:

```
patches:
  - type: FromCompositeFieldPath
```

```
fromFieldPath: spec.parameters.size
toFieldPath: spec.forProvider.settings.dataDiskSizeGb
transforms:
- type: math
  math:
    multiply: 4
```

After updating the composition, you will see two revisions. Only the latest revision will have the `current` flag of `true`. Also, we can notice that the MySQL provisioned with an automated revision update policy would have increased the storage. The following screenshot summarizes the output after applying the updated composition:

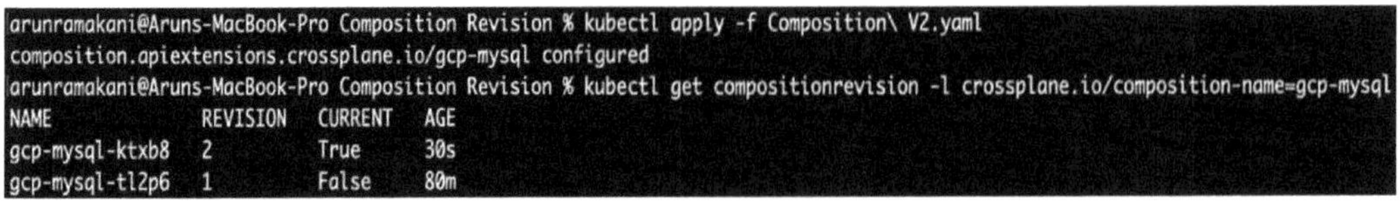

```
arunramakani@Aruns-MacBook-Pro Composition Revision % kubectl apply -f Composition\ V2.yaml
composition.apiextensions.crossplane.io/gcp-mysql configured
arunramakani@Aruns-MacBook-Pro Composition Revision % kubectl get compositionrevision -l crossplane.io/composition-name=gcp-mysql
NAME             REVISION   CURRENT   AGE
gcp-mysql-ktxb8  2          True      30s
gcp-mysql-tl2p6  1          False     80m
```

Figure 5.4 – New Composition Revision

Finally, we can manually upgrade the second MySQL instance by adding the `spec.compositionRevisionRef` attribute to the XR/Claim configuration. The field will hold the autogenerated composition revision name. The composition revision hands-on journey example is available at `https://github.com/PacktPublishing/End-to-End-Automation-with-Kubernetes-and-Crossplane/tree/main/Chapter05/Hand-on-examples/Composition-Revision`. In the following section, we will explore the ways to change the XR API contract.

API contract changes

API implementation details are just one direction in which XR changes can evolve. The highly interoperable API contract between the XR creating and consuming teams also needs to change over time. Contract change can fall under two categories:

- **Non-breaking changes**: The XR API will be backward-compatible, meaning that consumers are either not impacted by the change or can choose to adopt the new changes at their phase.

- **Breaking changes**: The XR API will not be backward-compatible. A new API version must be introduced, and the old API version must be deprecated at an appropriate time. All old API users should be safely migrating to the new API version.

Let's delve into non-breaking changes.

Non-breaking changes

Adding one or more optional parameters to the XRD contract can be considered a non-breaking change. It is non-breaking because the old external resources provisioned can co-exist with the new schema as the new parameters are optional. Note that removing an existing optional parameter in the XRD is a breaking change as Crossplane upfront does not know how to reconcile existing provisioned resources. A simple way to think about this is that if Composition/CompositionRevision can handle the co-existence of old and newly provisioned resources, then the XRD contract change is non-breaking.

A new optional parameter in the MySQL XR to choose the disk size is an example of a non-breaking change. The change will involve both a contract change and a composition revision. Let's go through a hands-on journey to make the previous XR example. All the configuration YAML required for this journey is available at `https://github.com/ PacktPublishing/End-to-End-Automation-with-Kubernetes-and-Crossplane/tree/main/Chapter05/Hand-on-examples/XRD-Contract-Change-Non-Breaking`. Refer to the following screenshot of our hands-on journey:

```
arunramakani@Aruns-MacBook-Pro XRD Contract Change - Breaking % kubectl apply -f xrd\ v1.yaml
compositeresourcedefinition.apiextensions.crossplane.io/xmysqls.xrd-non-breaking.imarunrk.com created
arunramakani@Aruns-MacBook-Pro XRD Contract Change - Breaking % kubectl apply -f Composition\ V1.yaml
composition.apiextensions.crossplane.io/gcp-mysql created
arunramakani@Aruns-MacBook-Pro XRD Contract Change - Breaking % kubectl apply -f Claim\ v1.yaml
mysql.xrd-non-breaking.imarunrk.com/mysql-db-v1 created
arunramakani@Aruns-MacBook-Pro XRD Contract Change - Breaking % kubectl get composition
NAME        AGE
gcp-mysql   67s
arunramakani@Aruns-MacBook-Pro XRD Contract Change - Breaking % kubectl get compositionrevision
NAME             REVISION   CURRENT   AGE
gcp-mysql-wphhg  1          True      88s
arunramakani@Aruns-MacBook-Pro XRD Contract Change - Breaking % kubectl get xrd
NAME                                    ESTABLISHED   OFFERED   AGE
xmysqls.xrd-non-breaking.imarunrk.com   True          True      2m
arunramakani@Aruns-MacBook-Pro XRD Contract Change - Breaking % kubectl get mysql -n alpha
NAME         READY   CONNECTION-SECRET   AGE
mysql-db-v1  True                        5m25s
arunramakani@Aruns-MacBook-Pro XRD Contract Change - Breaking % kubectl apply -f xrd\ v2.yaml
compositeresourcedefinition.apiextensions.crossplane.io/xmysqls.xrd-non-breaking.imarunrk.com configured
arunramakani@Aruns-MacBook-Pro XRD Contract Change - Breaking % kubectl apply -f Composition\ V2.yaml
composition.apiextensions.crossplane.io/gcp-mysql configured
arunramakani@Aruns-MacBook-Pro XRD Contract Change - Breaking % kubectl get composition
NAME        AGE
gcp-mysql   6m58s
arunramakani@Aruns-MacBook-Pro XRD Contract Change - Breaking % kubectl get compositionrevision
NAME             REVISION   CURRENT   AGE
gcp-mysql-6kmtv  2          True      13s
gcp-mysql-wphhg  1          False     7m6s
arunramakani@Aruns-MacBook-Pro XRD Contract Change - Breaking % kubectl apply -f Claim\ v2.yaml
mysql.xrd-non-breaking.imarunrk.com/mysql-db-v2 created
arunramakani@Aruns-MacBook-Pro XRD Contract Change - Breaking % kubectl get mysql -n alpha
NAME         READY   CONNECTION-SECRET   AGE
mysql-db-v1  True                        13m
mysql-db-v2  True                        6m28s
```

Figure 5.5 – Non-breaking contract change

The following are the steps to be performed throughout the hands-on journey to experiment with the non-breaking contract change:

1. Create the first version of the XRD in the target cluster (`xrd-v1.yaml`). The schema has `vm` as a mandatory parameter.

2. Create the first revision of the composition (`Composition-V1.yaml`). It will patch the `vm` value back into the `MR-CloudSQLInstance` tier attribute.

3. Now, the MySQL resource can be provisioned with `db-n1-standard-1` as the tire in GCP (`Claim-v1.yaml`).

4. Update and apply the XRD with an additional optional parameter, `size`, to specify the database disk size (`xrd-v2.yaml`).

5. Update and apply the new composition (`Composition-V2.yaml`). It will patch the additional size parameter into the MR.

6. Finally, create the second MySQL instance with a specific disk size and tire (`Claim-v2.yaml`).

7. To validate whether the first MySQL instance can be sill updated, change the tier with an update YAML (`Claim-v1-validate.yaml`).

We did not upgrade the API version when updating the contract. We will discuss this more in the upcoming section.

Version upgrade

In the previous section, we did not change the XRD version number from v1. Crossplane does not currently support XR version upgrades once a contract changes. API versioning without a contract change will be helpful in indicating API stability (alpha, beta, v1, and so on). We can just move from alpha to beta to a more stable version without changing the contract. The version upgrade is currently achieved by listing the old and new version definitions in the XRD. The `versions` array is the construct used for listing multiple versions. The two critical Boolean attributes under each version are `served` and `referenceable`. The `referenceable` flag will determine whether we can define a composition implementation for the given version. Only one version can have the referenceable flag set to `true`. This will be the version used by any new XR create/update event. A create/update event triggered by the old API version will still use the composition from the latest version, marked as referenceable. The `served` flag will indicate whether the given XR API version is in use. Some teams may still use the old version to consume the API. Switching off the `served` flag means that the given version is no longer available for clients. It will be the last step before removing the old version from the XR.

Look at a sample XRD with three versions, alpha, beta, and v1, at `https://github.`
`com/PacktPublishing/End-to-End-Automation-with-Kubernetes-`
`and-Crossplane/blob/main/Chapter05/Samples/XRD-Versions/`
`xrd-multiple-version.yaml`. This XRD has three versions. Version alpha will
no longer be served, and beta will be served but cannot be referred for resource creation
or update. The latest version, v1, will be the preferred version for any resource creation
or updates.

Kubernetes CRDs support multiple API versions both with and without an API contract
change. When there is an API contract change, a conversion webhook is configured by the
CRD author to support conversion between the versions. Conversions are required as CR
objects will be stored in the etcd with both old and new contracts. XRD, the Crossplane
equivalent to CRDs, does not take this approach. A conversion webhook involves
programming. Taking that route will violate the no-code agenda of Crossplane when
composing APIs. It's important to note that the Crossplane community is actively working to
build a configuration-based solution to support conversion and migration between versions.

Version upgrade with breaking changes

An alternative approach supports breaking contracts by introducing a new XR API
parallel. This approach uses an external naming technique and deletion policy to handle
breaking changes. With this pattern, we will migrate the resources to a new XR API and
remove the old API once the migration is finished in its entirety. The steps to achieve such
a version upgrade are as follows:

1. Create the v1 version of XRD. In the composition, define a standard nomenclature
    for naming the external resources (MRs). We should be able to reconstruct the
    names again in the new API. Generally, we can concatenate the XR and the
    composition name (`<XR>+'-'+<Composition>`). You can come up with
    a resource naming strategy that suits your environment. Maybe we can even use the
    namespace name to represent the product owner of the resource.

2. Ensure that all the MRs in the composition have `spec.deletionPolicy` defined
    as `Orphan`.

3. Let's say we have a couple of consumers for the XR, and they have created a few
    external resources. Assume that we have a policy requirement that requires breaking
    changes to the API contract.

4. To support the breaking change, delete all v1 versions of the XR. It will just delete
    the Crossplane references. The external resources are not deleted on account of the
    orphan deletion policy.

5. Then, delete the v1 version of XRD and Composition.

6. Finally, create the new v2 version of XRD with the same XR name. Update the composition to handle the recent breaking changes. Ensure that the new composition follows the same external resource name creation logic and maps to the new XRD version.

7. Create the deleted XR objects again, pointing to the v2 version of the API. The new XR objects will reference the old orphaned external resources. Crossplane controllers will reconcile any attribute value change.

Note that this type of migration to a new API version must be coordinated with all the XR-consuming teams. Once the migration is completed, the old API version will no longer be available. The following figure represents the migration process:

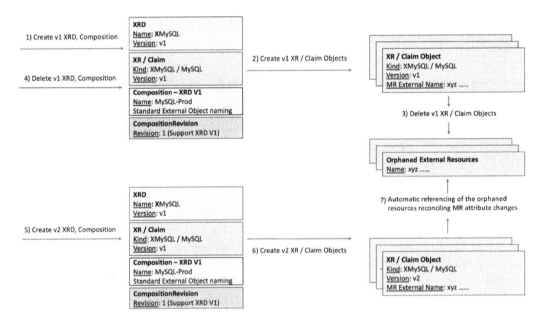

Figure 5.6 – Version migration

> **Tip**
> It is always good to have a standard way of generating external resource names. In addition to version migration, a reproducible naming pattern can afford several other advantages. Using pre-provisioned resources for a shared or cached infrastructure is an example of using the standard external resource naming pattern. Migrating resources to a new Crossplane environment can be another example.

It's recommended to go through a hands-on journey of breaking API contract, with the sample configuration provided at `https://github.com/PacktPublishing/End-to-End-Automation-with-Kubernetes-and-Crossplane/tree/main/Chapter05/Hand-on-examples/XRD-Contract-Change-Breaking`. Perform the following steps to go through the hands-on journey to handle breaking contract changes:

1. First, execute `xrd-v1.yaml`, `Composition-V1.yaml`, and `Claim.yaml`.

2. It will create an XRD and a Composition with the database size as optional parameter and the VM as a mandatory parameter. The Claim will provision the database wih the specified size and VM. The provisioned resource will get a standard external resource name.

3. Note that the `claim.name` label in `Claim.yaml` is used for constructing the external resource name in the composition section. It should be unique for every XR/Claim object to generate unique external resource names.

4. Let's now delete the v1 version of Claim, Composition, and XRD. When we delete the v1 claim, the external resource will not be deleted because the deletion policy is configured as an orphan.

5. Finally, apply the v2 version of Claim, Composition, and XRD. In the v2 XRD, we have broken the contract by removing the mandatory parameter, vm. The new v2 claim (`Claim-migrate.yaml`) will not have the vm parameter. Note that both Composition and Claim will point to the v2 version XRD.

6. Notice that the Crossplane will reclaim the orphaned resource and reconcile the virtual machine with the new default value provided in the Composition. We can validate that by looking into the GCP console or the Claim resource description.

Refer to the following screenshot where the preceding example is tested:

```
arunramakani@Aruns-MacBook-Pro XRD Contract Change - Breaking % kubectl apply -f xrd\ v1.yaml
compositeresourcedefinition.apiextensions.crossplane.io/xmysqls.xrd-breaking.imarunrk.com created
arunramakani@Aruns-MacBook-Pro XRD Contract Change - Breaking % kubectl apply -f Composition\ V1.yaml
composition.apiextensions.crossplane.io/gcp-mysql created
arunramakani@Aruns-MacBook-Pro XRD Contract Change - Breaking % kubectl apply -f Claim.yaml
mysql.xrd-breaking.imarunrk.com/my-db created
arunramakani@Aruns-MacBook-Pro XRD Contract Change - Breaking % kubectl get mysql -n alpha
NAME    READY   CONNECTION-SECRET   AGE
my-db   False                       44s
arunramakani@Aruns-MacBook-Pro XRD Contract Change - Breaking % kubectl get mysql -n alpha
NAME    READY   CONNECTION-SECRET   AGE
my-db   True                        5m6s
arunramakani@Aruns-MacBook-Pro XRD Contract Change - Breaking % kubectl delete mysql -n alpha
error: resource(s) were provided, but no name was specified
arunramakani@Aruns-MacBook-Pro XRD Contract Change - Breaking % kubectl delete mysql my-db -n alpha
mysql.xrd-breaking.imarunrk.com "my-db" deleted
arunramakani@Aruns-MacBook-Pro XRD Contract Change - Breaking % kubectl delete -f Composition\ V1.yaml
composition.apiextensions.crossplane.io "gcp-mysql" deleted
arunramakani@Aruns-MacBook-Pro XRD Contract Change - Breaking % kubectl delete -f xrd\ v1.yaml
compositeresourcedefinition.apiextensions.crossplane.io "xmysqls.xrd-breaking.imarunrk.com" deleted
arunramakani@Aruns-MacBook-Pro XRD Contract Change - Breaking % kubectl apply -f xrd\ v2.yaml
compositeresourcedefinition.apiextensions.crossplane.io/xmysqls.xrd-breaking.imarunrk.com created
arunramakani@Aruns-MacBook-Pro XRD Contract Change - Breaking % kubectl apply -f Composition\ V2.yaml
composition.apiextensions.crossplane.io/gcp-mysql created
arunramakani@Aruns-MacBook-Pro XRD Contract Change - Breaking % kubectl apply -f Claim\ migrate.yaml
mysql.xrd-breaking.imarunrk.com/my-db created
arunramakani@Aruns-MacBook-Pro XRD Contract Change - Breaking % kubectl get mysql -n alpha
NAME    READY   CONNECTION-SECRET   AGE
my-db   True                        37s
```

Figure 5.7 – XRD breaking changes

Following is the code snippet relating to external resource name patching from the preceding Composition example. This must be present in both composition versions, and the name generated should be the same for both versions:

```
- type: FromCompositeFieldPath
    fromFieldPath: metadata.name
    toFieldPath: metadata.annotations[crossplane.io/external-
name]
    transforms:
    - type: string
      string:
        fmt: "%s-gcp-mysql-cloudsqlinstance"
```

Note that we have used a new transform type to format the string before we patch. With this, we conclude the different ways of evolving the XR APIs. We will dive into an interesting case in the following section to build one XR composing another XR.

Nested and multi-resource XRs

Every software product depends on more than one infrastructure resource. It is essential to build single infrastructure recipes in order for the product teams to consume with a unified experience. The orchestration of infrastructure dependencies should remain abstracted. Such recipes require multiple resources to be composed into a single XR. In all the examples hitherto, we have always composed a single GCP resource inside an XR. Let's look at an XR sample where multiple GCP resources are composed into a single XR API. The following figure represents the resources and XR APIs that we are going to work with in the example:

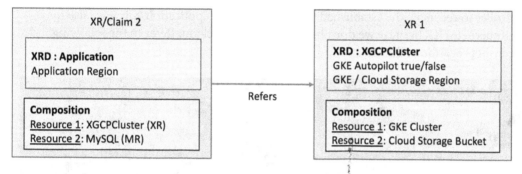

Figure 5.8 – Multi-resource nested XR

In addition to multiple resource provisioning in a single XR, we also have a nested XR pattern in *Figure 5.8*. We are composing three resources within two XRs. The first XR composes two resources, and the second XR composes the first XR and a database resource. Let's look at the details of each XR:

- **XR 1**: We will compose a Google Kubernetes Engine cluster and a Google Cloud storage resource with this XR. The idea is to provide cloud storage to hold the application logs for future analysis. Note that this XR will not have a claim name in the XRD definition. It will be of a cluster scope and a private API for the platform team. Product teams with access only to a namespace will not use this API directly. This XR will expose the region and the autopilot configuration as parameters. The region will be propagated back into both the resources and the autopilot configuration is used for the Kubernetes provisioning.

- **XR/Claim 2**: The second XR will compose the MySQL database, an MR, and the first XR to create a nested API. We will patch region parameters to the MySQL MR and propagate the same into the inner XR.

All examples in this hands-on journey are available at `https://github.com/PacktPublishing/End-to-End-Automation-with-Kubernetes-and-Crossplane/tree/main/Chapter05/Hand-on-examples/Nested-Multi-Resource-XR`.

Let's first create the XRD and Composition for both the XRs. Apply `xrd k8s.yaml`, `Composition k8s.yaml`, `xrd Application.yaml`, and `Composition Application.yaml` to the Crossplane cluster. You will see that the ESTABLISHED flag is True for both the XRDs. This indicates that the Crossplane has started a new controller to reconcile the established XR. The OFFERED flag will be True for the application XR and False for the Kubernetes XR. This indicates that the Crossplane has started a new controller to reconcile the established Claim only for the application XR. It is false for the Kubernetes XR because we don't have the respective claim. Refer to the following screenshot regarding XRD creation:

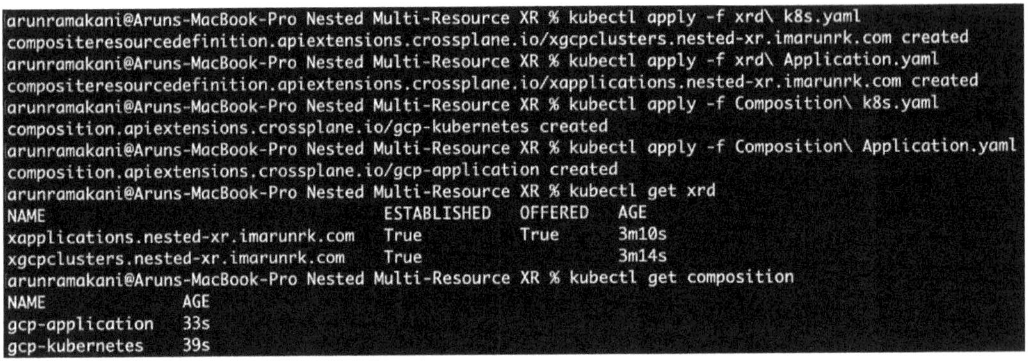

```
arunramakani@Aruns-MacBook-Pro Nested Multi-Resource XR % kubectl apply -f xrd\ k8s.yaml
compositeresourcedefinition.apiextensions.crossplane.io/xgcpclusters.nested-xr.imarunrk.com created
arunramakani@Aruns-MacBook-Pro Nested Multi-Resource XR % kubectl apply -f xrd\ Application.yaml
compositeresourcedefinition.apiextensions.crossplane.io/xapplications.nested-xr.imarunrk.com created
arunramakani@Aruns-MacBook-Pro Nested Multi-Resource XR % kubectl apply -f Composition\ k8s.yaml
composition.apiextensions.crossplane.io/gcp-kubernetes created
arunramakani@Aruns-MacBook-Pro Nested Multi-Resource XR % kubectl apply -f Composition\ Application.yaml
composition.apiextensions.crossplane.io/gcp-application created
arunramakani@Aruns-MacBook-Pro Nested Multi-Resource XR % kubectl get xrd
NAME                                    ESTABLISHED   OFFERED   AGE
xapplications.nested-xr.imarunrk.com    True          True      3m10s
xgcpclusters.nested-xr.imarunrk.com     True                    3m14s
arunramakani@Aruns-MacBook-Pro Nested Multi-Resource XR % kubectl get composition
NAME              AGE
gcp-application   33s
gcp-kubernetes    39s
```

Figure 5.9 – Nested XR-XRD and Composition

> **Tip**
> Similar to creating an XR API with multiple resources from a single cloud provider, we can also mix and match resources from multiple clouds. We just have to add the resources concerned with respective ProviderConfig clouds.

It's now time to create an application Claim resource. Apply Claim `Application.yaml` to the Crossplane cluster. You will see that a `CloudSQLInstance` instance, a cluster, and a bucket resource have been provisioned. Refer to the following screenshot where the resources are provisioned successfully:

```
arunramakani@Aruns-MacBook-Pro Nested Multi-Resource XR % kubectl apply -f Claim\ Application.yaml
application.nested-xr.imarunrk.com/my-application created
arunramakani@Aruns-MacBook-Pro Nested Multi-Resource XR % kubectl get application -n alpha
NAME              READY   CONNECTION-SECRET   AGE
my-application    False                       29s
arunramakani@Aruns-MacBook-Pro Nested Multi-Resource XR % kubectl get application -n alpha
NAME              READY   CONNECTION-SECRET   AGE
my-application    True                        28m
arunramakani@Aruns-MacBook-Pro Nested Multi-Resource XR % kubectl get Xapplication
NAME                    READY     COMPOSITION          AGE
my-application-tsjd9    True      gcp-application  28m
arunramakani@Aruns-MacBook-Pro Nested Multi-Resource XR % kubectl get CloudSQLInstance
NAME                          READY   SYNCED   STATE      VERSION     AGE
my-application-tsjd9-rgbst    True    True     RUNNABLE   MYSQL_5_7   28m
arunramakani@Aruns-MacBook-Pro Nested Multi-Resource XR % kubectl get XGCPCluster
NAME                          READY   COMPOSITION      AGE
my-application-tsjd9-ptv66    True    gcp-kubernetes   29m
arunramakani@Aruns-MacBook-Pro Nested Multi-Resource XR % kubectl get Cluster
NAME                          READY   SYNCED   STATE     ENDPOINT         LOCATION      AGE
my-application-tsjd9-js2h6    True    True     RUNNING   35.188.188.55    us-central1   29m
arunramakani@Aruns-MacBook-Pro Nested Multi-Resource XR % kubectl get Bucket
NAME                          READY   SYNCED   STORAGE_CLASS   LOCATION   AGE
my-application-tsjd9-zwrw4    True    True     STANDARD        US         29m
arunramakani@Aruns-MacBook-Pro Nested Multi-Resource XR % ▉
```

Figure 5.10 – Resource provisioning

If you would like to explore each resource in detail, use the Resource references. Execute
`kubectl describe application my-application -n alpha` to see the
details of the claim. It will refer to the XApplication XR object. If we look at the details
of the XApplication object, it will hold the reference to the CloudSQLInstance
MR and XGCPCluster XR. Similarly, we can go on till you reach the last MR. This is
beneficial for debugging activities. Sometimes you may see that the resources are not
getting ready. In those instances, explore each nested resource and refer to the events
section to ascertain what is happening. An example of referring nested resources from the
resource description is as follows:

```
Resource Ref:
  API Version:  nested-xr.imarunrk.com/v1
  Kind:         XApplication
  Name:         my-application-tsjd9
```

Figure 5.11 – Nested resource reference example 1

The preceding screenshot represented the Application claim description referring to the XApplication XR resource. The following screenshot represents the XApplication XR description referring to the XGCPCluster XR instance and CloudSQLInstance MR:

```
Resource Refs:
  API Version:    nested-xr.imarunrk.com/v1
  Kind:           XGCPCluster
  Name:           my-application-tsjd9-ptv66
  API Version:    database.gcp.crossplane.io/v1beta1
  Kind:           CloudSQLInstance
  Name:           my-application-tsjd9-rgbst
tatus:
```

Figure 5.12 – Nested resource reference example 2

The following is an example event that tells us that we have provided the wrong region as a parameter:

```
Spec:
  Deletion Policy:  Delete
  For Provider:
    Database Version:  MYSQL_5_7
    Region:            us-central1-b
    Settings:
      Tier:  db-g1-small
      User Labels:
        Crossplane - Kind:          cloudsqlinstance_database_gcp_crossplane_io
        Crossplane - Name:          my-application-cjgxj-zw7d9
        Crossplane - Providerconfig: gcp-credentials-project-1
  Provider Config Ref:
    Name:  gcp-credentials-project-1
Status:
  At Provider:
  Conditions:
    Last Transition Time:  2022-01-15T04:55:58Z
    Message:               create failed: cannot create new CloudSQL instance: googleapi: Error 400: Invalid request: Invalid value for r
egion: us-central1-b., invalid
    Reason:                ReconcileError
    Status:                False
    Type:                  Synced
Events:
  Type     Reason                          Age                    From                                                    Message
  ----     ------                          ----                   ----                                                    -------
  Warning  CannotInitializeManagedResource 2m52s                  managed/cloudsqlinstance.database.gcp.crossplane.io     cannot update Clo
udSQLInstance custom resource: Operation cannot be fulfilled on cloudsqlinstances.database.gcp.crossplane.io "my-application-cjgxj-zw7d9"
: the object has been modified; please apply your changes to the latest version and try again
  Warning  CannotCreateExternalResource    106s (x24 over 2m52s)  managed/cloudsqlinstance.database.gcp.crossplane.io     cannot create new
CloudSQL instance: googleapi: Error 400: Invalid request: Invalid value for region: us-central1-b., invalid
```

Figure 5.13 – Resource description with an error

> **Important**
>
> We need to follow many more patterns when we compose multiple resources to give a unified experience for product teams. The preceding example is a simple example to start the topic. We will see more on this in the upcoming chapters.

PatchSets

If you look at the composition in the preceding example, you can see that we have used a new pattern called **PatchSets**. If you find yourself repeating the same patch operation again and again under each resource, then PatchSets is the way to go. Here, we define the patch operation as a static function and include it under the required resource sections. The following is an example of the `patchSets` function definition to patch a region:

```
patchSets:
  - name: region
    patches:
    - type: FromCompositeFieldPath
      fromFieldPath: spec.parameters.region
      toFieldPath: spec.forProvider.region
```

We can define multiple `patchSet` functions. To include a specific patch set function within a given resource, use the following code snippet:

```
patches:
  - type: PatchSet
    patchSetName: region
```

We will see more nested and multi-resource XR examples in the upcoming chapters. In the following section, we will look at detailed configuration options for defining the XRD schema.

XRD detailed

While looking at **Composite Resource Definition** (**XRD**) in the previous chapter, we touched on limited configuration options required to learn the basics of XR. It's now time to look at more detailed configuration options to build clean and robust XR APIs. A significant part of the details we will look at are about openAPIV3Schema, which is used to define the input and output of the XR API. The following are the topics we will cover in this section:

- Naming the versions

- The openAPIV3Schema structure

- The additional parameter of an attribute

- Printer columns

Let's start with the *Naming the versions section*.

Naming the versions

The version name of our XRD cannot have any random string. It has a specific validation inherited from the CRDs and standard Kubernetes APIs. The string can contain only lowercase alphanumeric characters and -. Also, it must always start with an alphabetic character and end with an alphanumeric character, which means that - cannot be the start or ending character. Also, a number cannot be the starting character. Some valid versions are my-version, version-1, abc-version1, and v1. While we can have many permutations and combinations for naming a version, some standard practices are followed across CRDs. Following the same with XRDs will enable API consumers to understand the stability of the API. The version string starts with v followed by a number with these standards (v1, v2). This is then optionally followed by either alpha or beta, depending on the API's stability. Generally, the alpha string represents the lowest stability (v5alpha), while beta is the next stability level (v3beta). If both texts are missing, the XR is ready for production use. An optional number can follow the optional alpha/beta text representing the incremental releases (v2alpha1, v2alpha2, and so on).

If you have an invalid version string provided with the XRD, you will see that the XRD will not get configured properly. The ESTABLISHED flag will not be set to True. You apply the - xrd\ invalid\ version\ test.yaml file from the samples folder to see what happens when you have an incorrect version number. Refer to the following screenshot:

```
arunramakani@Aruns-MacBook-Pro XRD Versions % kubectl apply -f xrd\ invalid\ version\ test.yaml
compositeresourcedefinition.apiextensions.crossplane.io/xbuckets.version-test.imarunrk.com created
arunramakani@Aruns-MacBook-Pro XRD Versions % kubectl get xrd
NAME                                      ESTABLISHED    OFFERED    AGE
xbuckets.version-test.imarunrk.com                                  7s
```

Figure 5.14 – Invalid version XRD

Also, you will be able to see the following error logs in the Crossplane pod in the crossplane-system namespace:

```
2022-01-15T20:06:46.217Z       ERROR      crossplane.controller-
runtime.manager.controller.defined/compositeresourcedefinition.
apiextensions.crossplane.io      Reconciler error
{"reconciler group": "apiextensions.crossplane.io", "reconciler
kind": "CompositeResourceDefinition", "name": "xbuckets.
version-test.imarunrk.com", "namespace": "", "error": "cannot
apply rendered composite resource CustomResourceDefinition:
cannot create object: CustomResourceDefinition.apiextensions.
k8s.io \"xbuckets.version-test.imarunrk.com\" is invalid:
[spec.versions[0].name: Invalid value: \"v1.0\": a DNS-1035
```

```
label must consist of lower case alphanumeric characters or
'-', start with an alphabetic character, and end with an
alphanumeric character (e.g. 'my-name', or 'abc-123', regex
used for validation is '[a-z]([-a-z0-9]*[a-z0-9])?'), spec.
version: Invalid value: \"v1.0\": a DNS-1035 label must consist
of lower case alphanumeric characters or '-', start with an
alphabetic character, and end with an alphanumeric character
(e.g. 'my-name', or 'abc-123', regex used for validation is
'[a-z]([-a-z0-9]*[a-z0-9])?')]"}
```

> **Tip**
>
> When we troubleshoot an issue with Crossplane, logs from the Crossplane pod
> can help. Enable debugging mode by adding an argument, --debug, to the
> Crossplane pod. Similarly, we can even look at the provider's container logs.

The openAPIV3Schema structure

The specification of the XR API is defined using openAPIV3Schema. Every
configuration element in the XRD under this section represents the input and output of
the XR API:

```
versions:
  - name: v1alpha
    schema:
      openAPIV3Schema:
      # Input and output definition for the XR
```

Generally, we configure the openAPIV3Schema section with two objects, spec and
status. The spec object represents the API input, while the status object represents
the response. We can skip defining the status section in XRD if we don't have any
custom requirements. Crossplane would inject the standard status fields into the XR/
Claim. Refer to the following code snippet representing the openAPIV3Schema
configuration template for the XR API input and output:

```
openAPIV3Schema:
  type: object
  properties:
    # spec - the API input configuration
    spec:
      type: object
```

```
      properties:
        ............. configuration continues
    # status - the API output configuration
    status:
      type: object
      properties:
        ............ configuration continues
```

The schema configuration is all about a mix of - attributes, their types, and properties. An attribute type of object will hold a list of properties. For example, the root attribute openAPIV3Schema: is of the object type followed by a list of properties (spec and status). A list of properties is nothing but a list of attributes. Suppose the attribute type is primitive, such as string or integer. Such an attribute will be the end node. The object-properties recursion can continue in as much depth as we require. Refer to the following code snippet:

```
# The root attribute openAPIV3Schema of type object
openAPIV3Schema:
  type: object
  # spec/status - attributes (properties) of openAPIV3Schema
  properties:
    # spec - the XR input
    spec:
      type: object
      properties:
        # parameters - again an object with attributes list
        parameters:
          type: object
          properties:
            # region - string primitive  - node ends
            region:
              type: string
    # status - API output configuration
    # The exact structure of configuration as before
    # Attributes, their types, and properties
    status:
      type: object
```

```
        properties:
          zone:
            description: DB zone.
            type: string
```

In the following section, we can look at a few additional valuable configuration options along with the basic openAPIV3Schema configuration.

The additional parameter of an attribute

The attribute node can configure a few other critical configurations that API developers will use daily. Following are some of the frequently used configurations:

- **Description** is a string that will help us provide valuable information for the API consumers about the attribute. It can hold information about the use of the parameter, possible values we can configure, and validation requirements.

- **Required** is an attribute representing the list of mandatory inputs that are required from the user for the API.

- **Default** is an attribute that provides a default value if the user does not input a value.

- **Enum** can configure the list of possible values for a given attribute.

In addition to these fields, there is a list of validation-related configurations including minimum, maximum, pattern, maxLength, and minLength. Refer to the following sample configuration:

```
spec:
  type: object
  description: API input specification
  properties:
    parameters:
      type: object
      description: Parameter's to configure the resource
      properties:
        size:
          type: integer
          description: Disk size of the database
          default: 20
```

```
        minimum: 10
        maximum: 100
      vm:
        type: string
        description: Size of the virtual machine.
      enum:
      - small
      - medium
      - large
    required:
    - size
  required:
  - parameters
```

To explore more detailed possibilities, visit `https://github.com/OAI/OpenAPI-Specification/blob/main/versions/3.0.0.md#schemaObject`.

> **Tip**
> We can use the description field to announce the parameter deprecation information. This technique can be helpful in delaying breaking changes to a contract by making a mandatory field optional with a deprecation message.

Printer columns

We can use the printer columns to add what `kubectl` will display when we get the resource list. We should provide a name, data type, and JSON path mapping to the attribute we wish to display for each column. Optionally, we may also provide a description. Refer to the following sample configuration:

```
additionalPrinterColumns:
- name: Zone
  type: string
  description:
  jsonPath: .spec.zone
- name: Age
  type: date
  jsonPath: .metadata.creationTimestamp
```

The printer column configuration remains parallel to the schema configuration.

This concludes our discussion of detailed XRD configuration. We have covered most of the configuration required for day-to-day work, but there are endless possibilities. It will add value by reading up on CRD at `https://kubernetes.io/docs/tasks/extend-kubernetes/custom-resources/custom-resource-definitions/`.

Managing external software resources

We have always talked about managing external infrastructure resources using Crossplane from the beginning of this book. However, it does not always have to be just an infrastructure resource. We could even manage external software applications from Crossplane. For a software application to be able to work best with the Crossplane ecosystem, it must have the following qualities:

- We should have well-defined and stable APIs to perform CRUD operations.

- The API should have a high-fidelity design with filters to control granular application configuration.

It's time to look at an example. Think about deploying an application in Kubernetes using Helm. Helm can package any application and provide a well-defined CRUD API to deploy, read, update, and uninstall. Above all, we can create granular control over the application configuration with parameters. We have a helm Crossplane provider already available and used extensively by the community. The idea of managing external applications from a Crossplane control plane can enable a new world of unifying application and infrastructure automation. The following section will cover the unifying aspect in more detail.

Unifying the automation

Managing external software resources with Crossplane is the crossroad for unifying infrastructure and application DevOps. We could package software and infrastructure dependencies into a single XRs. Such a complete package of applications and infrastructure introduces numerous advantages, some of which are listed here:

- The approach will unify the tooling and skills required for application and infrastructure automation.

- More importantly, the entire stack will enjoy the advantages of the Kubernetes operating model.

- Integrating vendor software into an enterprise ecosystem will become quicker and more standardized. Software vendors can quickly build packages that fit into different ecosystems. Currently, software vendors must custom-build for the individual cloud provider marketplace. This approach can assist in building a universal vendor software marketplace.

- We can easily apply the audit process to comply with any compliance standards. Previously, this would have been complicated as software and its infrastructure dependencies are spread about.

The following figure represents a unified XR API:

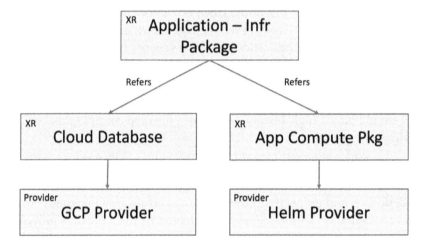

Figure 5.15 – Unified XR

> **Important**
>
> In a later chapter, we can go through a hands-on journey to experience building an XR API covering both applications and infrastructure dependencies.

Summary

I hope it's been fun to read this chapter and go through the hands-on journey. It covered different patterns that are useful in our day-to-day work when adopting Crossplane. We covered different ways to evolve our XR APIs, detailed XR configurations, how to manage application resources, and nested and multi-resource XRs. There are more patterns to be covered.

The next chapter will discuss more advanced Crossplane methods and their respective hands-on journeys.

6
More Crossplane Patterns

Following the previous chapter, we will continue to discover more Crossplane patterns that are key to building a state-of-the-art infrastructure automation platform. We will cover different topics, such as managing dependencies between resources, propagating secrets, using the Crossplane Helm provider, trade-off points in defining the XR API boundary, and monitoring the Crossplane control plane using Prometheus. Throughout the chapter, we will use examples with a hands-on journey to understand these concepts. We have been using GCP in all the previous chapters. In this chapter, we will use both GCP and AWS to learn Crossplane. Finally, we will learn more debugging skills, which are vital for day-to-day platform development and operations.

The following are the topics covered in the chapter:

- AWS provider setup
- Managing dependencies
- Secret propagation hands-on
- Helm provider hands-on
- Defining API boundaries
- Alerts and monitoring
- More troubleshooting patterns

AWS provider setup

Some of the examples in this chapter will use AWS as the cloud provider. Apart from GCP, we are covering AWS to establish what it takes to work with a new cloud provider. It will help us realize how working with one cloud provider will enable us to be competent enough to handle any cloud provider in Crossplane. We can look at the AWS provider setup in the following three steps:

1. Creating an AWS account and IAM user

2. Creating the Kubernetes secret

3. Provider and ProviderConfig setup

Creating an AWS account and IAM user

You can register with AWS and use some of its services free, provided you have a credit card. You can read more about the AWS free tier at `https://aws.amazon.com/free/free-tier-faqs/`. Once you have the free account created, the next step is to create a new IAM user. The following screenshots will cover the different stages in the IAM user creation. Go to the **IAM** section in the AWS web console and click **Add a user**. Select the credentials type as an access key shown in the following screenshot:

Set user details

You can add multiple users at once with the same access type and permissions. Learn more

User name*   crossplane-control-plane

⊕ Add another user

Select AWS access type

Select how these users will primarily access AWS. If you choose only programmatic access, it does NOT prevent users from accessing the console using an assumed role. Access keys and autogenerated passwords are provided in the last step. Learn more

Select AWS credential type*   ✔   **Access key - Programmatic access**
Enables an **access key ID** and **secret access key** for the AWS API, CLI, SDK, and

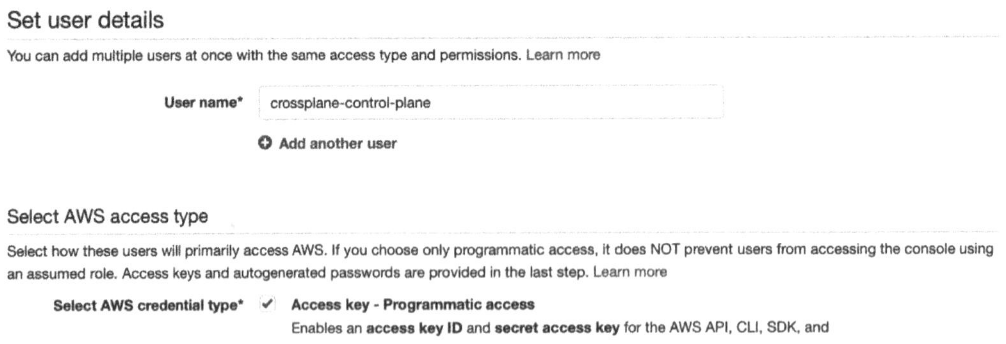

Figure 6.1 – Creating a user

The next step is to add the user to an access group. If you don't have a user group already, you can use the **Create group** button and create a new group with appropriate access control. Alternatively, we can attach an existing access policy to the user or copy permissions from a current user. Ensure that you have added the required permissions for the resources provisioned through Crossplane. I have provided an **AdministratorAccess** role to provide full access to all AWS resources.

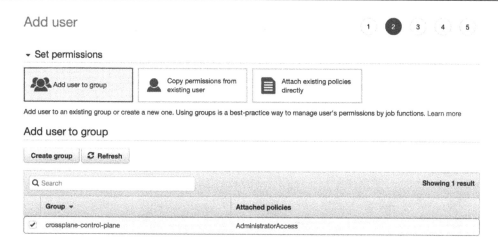

Figure 6.2 – Add user to group

Now you will have the access ID and secret of the new IAM user in the AWS console, which will be helpful for Crossplane AWS Provider configuration:

Figure 6.3 – New IAM user

The next step is to use the access key ID and the secret key of the IAM user to configure the Kubernetes secret.

Creating the Kubernetes secret

Creating the Kubernetes secret starts with setting up the AWS CLI in your local environment. It will be a simple step to download the installable and perform the installation. Follow the installation instructions at `https://docs.aws.amazon.com/cli/latest/userguide/getting-started-install.html`. Next, we can create the AWS login profile using the `aws configure --profile default` command. It will ask for the access key ID, secret access key, default region, and output format. The access key ID and the secret key are what we got while creating the IAM user. You can ignore the default region and output format.

```
arunramakani@Aruns-MacBook-Pro aws-setup % aws configure --profile default
AWS Access Key ID [****************HL7W]:
AWS Secret Access Key [****************w9HY]:
Default region name [us-west-2]:
Default output format [json]:
```

Figure 6.4 – Login profile

The next step is to create the Kubernetes secret using the preceding profile. Execute the following commands:

```
# Set a variable with the profile name
AWS_PROFILE=default
# Create a configuration file with profile data
echo -e "[$AWS_PROFILE]\naws_access_key_id = $(aws configure
get aws_access_key_id --profile $AWS_PROFILE)\naws_secret_
access_key = $(aws configure get aws_secret_access_key
--profile $AWS_PROFILE)" > aws-credentials.conf
# Create kubernetes secret from the configuration file
kubectl create secret generic aws-credentials -n crossplane-
system --from-file=creds=./aws-credentials.conf
```

Refer to the following screenshot where the Kubernetes secret is created:

```
arunramakani@Aruns-MacBook-Pro aws-setup % AWS_PROFILE=default
arunramakani@Aruns-MacBook-Pro aws-setup % echo -e "[$AWS_PROFILE]\naws_access_key_id = $(aws configure get aws_access_key_id --profile $
AWS_PROFILE)\naws_secret_access_key = $(aws configure get aws_secret_access_key --profile $AWS_PROFILE)" > aws-credentials.conf
arunramakani@Aruns-MacBook-Pro aws-setup % kubectl create secret generic aws-credentials -n crossplane-system --from-file=creds=./aws-cre
dentials.conf
secret/aws-credentials created
```

Figure 6.5 – Kubernetes secret creation

We are now done with the creation of Kubernetes secrets. The following section will look at the AWS provider installation and setup in the Crossplane environment.

AWS provider and ProviderConfig setup

Install Crossplane AWS provider by applying the following YAML to the cluster. The configuration has two parts to it. The provider configuration will install the AWS provider, and `ControllerConfig` enables debugging mode to the provider pod logs. It is not mandatory to have the `ControllerConfig` configuration. The example here will be helpful when you want to debug an issue. Note that the `ControllerConfig` name refers to the provider configuration:

```
apiVersion: pkg.crossplane.io/v1alpha1
kind: ControllerConfig
metadata:
  name: debug-config
spec:
  args:
    - --debug
---
```

```
apiVersion: pkg.crossplane.io/v1
kind: Provider
metadata:
  name: provider-aws
spec:
  package: "crossplane/provider-aws:v0.23.0"
  controllerConfigRef:
    name: debug-config
```

Finally, apply the following provider configuration YAML referring to the secret:

```
apiVersion: aws.crossplane.io/v1beta1
kind: ProviderConfig
metadata:
  name: aws-credentials
spec:
  credentials:
    source: Secret
    secretRef:
      namespace: crossplane-system
      name: aws-credentials
      key: creds
```

We are ready to create the resources from the AWS free tier and experiment. All the setup instructions are available at `https://github.com/PacktPublishing/End-to-End-Automation-with-Kubernetes-and-Crossplane/tree/main/Chapter06/Hand-on-examples/aws-setup`. Execute the `rds.yaml` file to validate whether the AWS provider setup is down proper. The following screenshot refers to the successful provisioning of an RDS resource from AWS:

```
arunramakani@Aruns-MacBook-Pro aws-setup % kubectl apply -f rds.yaml
rdsinstance.database.aws.crossplane.io/rdspostgresql created
arunramakani@Aruns-MacBook-Pro aws-setup % kubectl get RDSInstance
NAME            READY   SYNCED   STATE       ENGINE     VERSION   AGE
rdspostgresql   True    True     available   postgres   12.8      6m6s
```

Figure 6.6 – RDS provisioning

This completes the AWS setup activities. The following section will look at resource referencing to manage dependencies between the resources.

Managing dependencies

One external resource referencing another resource is a recurring pattern in infrastructure. For example, we may want to provision our Kubernetes cluster in a specific **Virtual Private Network (VPN)**. The S3 bucket policy definition referring to the S3 bucket is another example. We could go on with many such examples. From the perspective of building an XR API, there will be a requirement to establish dependencies between external resources within a given XR or in a nested XRs scenario, or between resources in independent XRs. Crossplane offers three different ways to refer one resource from another. Each of these options has its use case:

- **Direct reference**: This configuration option refers to the resources directly with a unique identifier such as a resource name or an **Amazon Resource Name (ARN)** or other identifier based on the specific cloud provider and resource type. For example, consider the AWS resource `UserPolicyAttachment`. It can attach an IAM user to a `Policy` object. Here, the reference to the `Policy` object can be done using the attribute called `PolicyARN` (ARN reference). Similarly, a `User` object reference can be executed using the `UserName` attribute (name reference).

- **Selector reference within the XR**: This option refers to the resources within the XR using a `selector` attribute. `selector` is an attribute that instructs Crossplane to look for the referring resources based on the conditions specified in its sub-attributes. `MatchControllerRef` and `MatchLabels` are the sub-attributes of the `selector` attribute. The `MatchControllerRef` value will be `true`, guiding Crossplane to look for the referring resources within the XR. The second attribute, `MatchLabels`, drives Crossplane to look for referring resources with the specified labels. If the selector identifies more than one recourse, one of the resources is selected randomly. If the direct reference attribute discussed in the previous option is present in the configuration, the `selector` attributes will be ignored.

- **Selector reference outside the XR**: Its behavior is the same as option two, excluding the false `MatchControllerRef` value. It guides the Crossplane to look for matching resources outside the XR.

> **Tip**
>
> We can use two strategies to identify the value for direct reference configuration. We can create the resources with a predictable name to reconstruct them again at the reference point. It is similar to what we discussed about external resource names in the last chapter. If the unique identifier is a cloud-generated ID such as ARN, copy the identifier to a custom-defined status attribute (XR API response) for usage at a later point in time.

Don't worry if it's confusing. Let's look at the resource reference with a couple of hands-on examples. The first example will cover the direct and selector configurations within and nested XR.

Resource reference within and nested XR

The example will be a real-world scenario. We will create an S3 bucket with a specific IAM policy and create an IAM user who can access the bucket. The following are the managed resources involved in the example:

- `Bucket`: This is an MR to create an S3 bucket. We will use this to provision the bucket in a specific region.

- `Policy`: This is the MR part of the AWS IAM resources list. It is helpful in defining usage guidelines for a given resource. In the example here, we will create a policy with full access to read and edit all objects in the bucket.

- `User`: The MR represents the AWS IAM user. We will create a new user to access the created bucket.

- `UserPolicyAttachment`: This is again part of the list of resources under AWS IAM. This MR will attach a policy to a user. We will link the bucket policy we created to the user.

You can see that there is a requirement for referring one resource from another. For example, a `Policy` resource would have to refer to the bucket name to build the policy configuration. Another example is `UserPolicyAttachment`, referring to the `Policy` and `User` resources to attach them. The following diagram will represent the relation between the resources, their reference option, and the XR boundary:

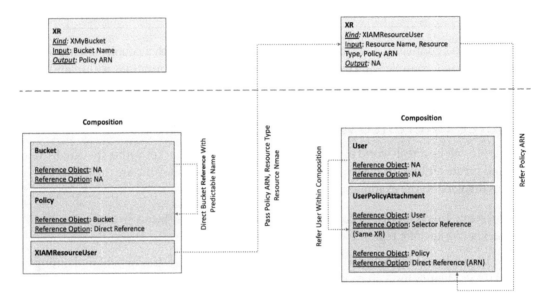

Figure 6.7 – Resource reference within and nested XR

The complete example with XRD, composition, and Claim is available at `https://github.com/PacktPublishing/End-to-End-Automation-with-Kubernetes-and-Crossplane/tree/main/Chapter06/Hand-on-examples/same-nested-xr-reference`. Let's look at some of the essential code snippets to get comfortable with the example and the concept. Bucket name reference within the policy document is the first snippet we will see. Both `Policy` and `Bucket` MRs are in the same composition. The requirement is to refer to the bucket ARN name within the policy document JSON. Thankfully ARN identifiers have a predictable format, and we can construct the ARN from the bucket's name. The bucket's name is already available as both MRs are in the same composition, and the bucket name is an XR API input. Following is the code snippet showing the resource reference discussed. It patches the policy document attribute using the `CombineFromComposite` patch type. Here, the bucket name is embedded directly using an `fmt` string operation:

```
- type: CombineFromComposite
  toFieldPath: spec.forProvider.document
```

```
combine:
  variables:
  - fromFieldPath: spec.parameters.bucketName
  - fromFieldPath: spec.parameters.bucketName
  strategy: string
  string:
    fmt: |
      {
          "Version": "2012-10-17",
          "Statement": [
            {
              "Effect": "Allow",
              "Action": [ "s3:*" ],
              "Resource": [
                "arn:aws:s3:::%s",
                "arn:aws:s3:::%s/*"
              ]
            }
          ]
      }
```

Next, we will look at how the `Policy` resource ARN is extracted to pass it on to the inner nested XR - `XIAMResourceUser`. It works in two steps:

1. Patch the ARN identifier of the `Policy` object back into the API response attribute.

2. Patch the extracted ARN identifier as an API input to the nested XR (`XIAMResourceUser`).

Note that initially, `XIAMResourceUser` will fail till the `Policy` object is wholly created and ARN is available. It is the typical control-plane behavior to make the resources eventually consistent. Following is the code snippet of the ARN patching from two resources, `Policy` and `XIAMResourceUser`:

```
# Policy - Patch API response with ARN
- type: ToCompositeFieldPath
  fromFieldPath: status.atProvider.arn
  toFieldPath: status.policyARN
```

```
# XIAMResourceUser - patch the policy arn as API input
- fromFieldPath: status.policyARN
  toFieldPath: spec.parameters.policyARN
```

Finally, we will look at the code snippet of the UserPolicyAttachment resource, where we have two external resources (User and Policy) using the different referencing methods. The policy reference will be made directly with the ARN identifier, and the user reference will be made using the selector. Refer to the following code:

```
- base:
  apiVersion: iam.aws.crossplane.io/v1beta1
  kind: UserPolicyAttachment
  spec:
    providerConfigRef:
      name: aws-credentials
    forProvider:
    # Selectors refer to the User from the same composition
      userNameSelector:
        matchControllerRef: true
  patches:
  # Patch the resource name
  # <Type>-<Parent Type>-<Parent Resource Name>
  - type: CombineFromComposite
    toFieldPath: metadata.name
    combine:
      variables:
      - fromFieldPath: spec.parameters.resourceType
      - fromFieldPath: spec.parameters.resourceName
      strategy: string
      string:
        fmt: "policy-attachement-%s-%s"
  # Patch the policy ARN reference
  - toFieldPath: spec.forProvider.policyArn
    fromFieldPath: spec.parameters.policyARN
```

To execute the example yourself and validate the references, follow the next steps:

1. Apply XRDs and compositions to the target Crossplane.

2. Next, apply the Claim configuration. It will create all the required resources and establish the required dependencies.

The following screenshot shows successful bucket creation in AWS:

Figure 6.8 – S3 bucket provisioned

The following screenshot shows all the execution steps of the example:

```
arunramakani@Aruns-MacBook-Pro same-nested-xr-reference % kubectl apply -f xrd-bucket.yaml
compositeresourcedefinition.apiextensions.crossplane.io/xmybuckets.reference.imarunrk.com created
arunramakani@Aruns-MacBook-Pro same-nested-xr-reference % kubectl apply -f xrd-IAM.yaml
compositeresourcedefinition.apiextensions.crossplane.io/xiamresourceusers.reference.imarunrk.com created
arunramakani@Aruns-MacBook-Pro same-nested-xr-reference % kubectl apply -f composition-bucket.yaml
composition.apiextensions.crossplane.io/my-bucket created
arunramakani@Aruns-MacBook-Pro same-nested-xr-reference % kubectl apply -f composition-IAM.yaml
composition.apiextensions.crossplane.io/resource-iam created
arunramakani@Aruns-MacBook-Pro same-nested-xr-reference % kubectl get xrd
NAME                                        ESTABLISHED   OFFERED   AGE
xiamresourceusers.reference.imarunrk.com    True                    27s
xmybuckets.reference.imarunrk.com           True          True      34s
arunramakani@Aruns-MacBook-Pro same-nested-xr-reference % kubectl get composition
NAME           AGE
my-bucket      25s
resource-iam   17s
arunramakani@Aruns-MacBook-Pro same-nested-xr-reference % kubectl apply -f claim-bucket.yaml
mybucket.reference.imarunrk.com/my-bucket created
arunramakani@Aruns-MacBook-Pro same-nested-xr-reference % kubectl get MyBucket -n alpha
NAME          READY   CONNECTION-SECRET   AGE
my-bucket     True                        6m19s
```

Figure 6.9 – Example execution

Also, note that the `User` object is created with the `Policy` resource attached in the AWS console:

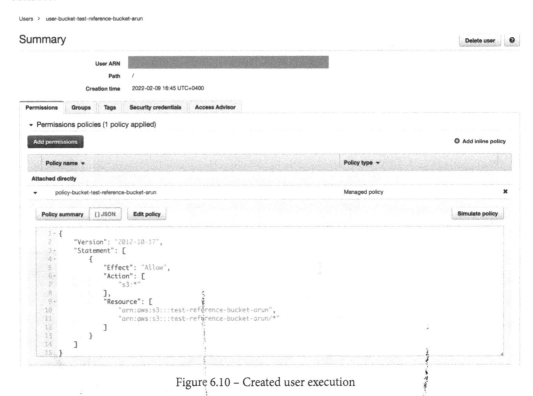

Figure 6.10 – Created user execution

Finally, refer to the screenshot showing the events where `XIAMResourceUser` fails owing to the unavailability of the policy ARN. It will work automatically once the policy ARN is available:

Figure 6.11 – XIAMResourceUser failure event

Information

Please note that we have not used `MatchLabels` in our selector reference. Only `MatchControllerRef` was used with `true` as the value. In this case, there was only one `User` object in the same composition, which can be referred to without any issue. We will use `MatchLabels` if we have more than one `User` object within the composition or if we want to refer to a recourse external to the current composition.

We are done with the exploration of referring resources within and nested XR. We will refer to a resource outside the composition in the following section.

Referring to an outside resource

To refer to a resource outside the composition, we will use `MatchLabels` and `MatchControllerRef`. `MatchControllerRef` should be specified as `false`. This would refer to an outside resource, MR, or another resource inside a Claim/XR. We will modify the last example into two independent XRs and ensure that the `UserPolicyAttachment` object can refer to the `Policy` object from an independent XR using label selectors. The following diagram will represent the relation between the resources, their reference option, and the XR boundary:

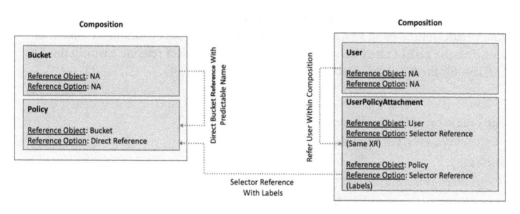

Figure 6.12 – Referring to an outside resource

Note that the XRs are not nested here. The `XMyBucket` XR will not have `XIAMResourceUser` as one of the resources. Providing a scenario where both XRs are independent, the `Policy` object must refer to the XR using a label selector. Let's look at a couple of essential code snippets that reference resources using selector labels. Following is the code that adds a couple of labels to the `Policy` resource. The first label, `resourceType`, is added directly to the metadata. The second label, `resourceName`, is patched using the bucket name, which is the input parameter for the XR:

```
- base:
    apiVersion: iam.aws.crossplane.io/v1beta1
```

```
kind: Policy
metadata:
  # Add labels one as the resource type
  labels:
    resourceType: bucket
spec:
  providerConfigRef:
    name: aws-credentials
  forProvider:
    path: "/"
patches:
# patch labels two from the resource name
- fromFieldPath: spec.parameters.bucketName
  toFieldPath: metadata.labels[resourceName]
```

The next part of the code will patch both `resourceName` and `resourceType` labels to the `UserPolicyAttachment` resource. It will be patched under policyArnSelector's `MatchLabels` attribute. Both label values are part of the XR API input. You can decide on your predictable labeling strategy to make this discovery process standard. Note that the `MatchControllerRef` value is `true` for the `User` object reference within the XR and `false` for the `Policy` object reference across the XR:

```
- base:
  apiVersion: iam.aws.crossplane.io/v1beta1
  kind: UserPolicyAttachment
  spec:
    providerConfigRef:
      name: aws-credentials
    forProvider:
      # refer to the IAM user from the same composition
      userNameSelector:
        matchControllerRef: true
      policyArnSelector:
        matchControllerRef: false
  patches:
  # Patch the policy ARN lable 1
  - toFieldPath: spec.forProvider.policyArnSelector.matchLabels.
resourceName
```

```
       fromFieldPath: spec.parameters.resourceName
   # Patch the policy ARN lable 2
   - toFieldPath: spec.forProvider.policyArnSelector.matchLabels.
 resourceType
       fromFieldPath: spec.parameters.resourceType
```

The example discussed is available at `https://github.com/PacktPublishing/End-to-End-Automation-with-Kubernetes-and-Crossplane/tree/main/Chapter06/Hand-on-examples/different-xr-reference`. To fully experience resource provisioning, apply `composition-IAM.yaml`, `composition-bucket.yaml`, `xrd-IAM.yaml`, and `xrd-bucket.yaml` to the target Crossplane cluster. It will create both XRs and their respective compositions. Then, apply `claim-bucket.yaml` and `claim-iam.yaml` to create the resources. The following screenshot covers full execution of the example:

```
arunramakani@Aruns-MacBook-Pro different-xr-reference % kubectl apply -f xrd-bucket.yaml
compositeresourcedefinition.apiextensions.crossplane.io/xmybuckets.differentxr.reference.imarunrk.com created
arunramakani@Aruns-MacBook-Pro different-xr-reference % kubectl apply -f xrd-IAM.yaml
compositeresourcedefinition.apiextensions.crossplane.io/xiamresourceusers.differentxr.reference.imarunrk.com created
arunramakani@Aruns-MacBook-Pro different-xr-reference % kubectl apply -f composition-bucket.yaml
composition.apiextensions.crossplane.io/my-bucket created
arunramakani@Aruns-MacBook-Pro different-xr-reference % kubectl apply -f composition-IAM.yaml
composition.apiextensions.crossplane.io/resource-iam created
arunramakani@Aruns-MacBook-Pro different-xr-reference % kubectl apply -f claim-bucket.yaml
mybucket.differentxr.reference.imarunrk.com/my-bucket created
arunramakani@Aruns-MacBook-Pro different-xr-reference % kubectl apply -f claim-iam.yaml
iamresourceuser.differentxr.reference.imarunrk.com/iam-user created
arunramakani@Aruns-MacBook-Pro different-xr-reference % kubectl get mybucket -n alpha
NAME        READY   CONNECTION-SECRET    AGE
my-bucket   True                         11m
arunramakani@Aruns-MacBook-Pro different-xr-reference % kubectl get IAMResourceUser -n alpha
NAME       READY   CONNECTION-SECRET    AGE
iam-user   True                         11m
```

Figure 6.13 – Referring to an outside resource – Example

Like *Figure 6.10*, the `User` object will be created with the `Policy` resource attached in the AWS console. We have now completed our exploration of resource references. The following section will look at secret propagation with a hands-on example.

Secret propagation hands-on

Secret propagation is a critical Crossplane pattern, as all resources provisioned generally require credentials to access the resource. We covered the same topic in *Chapter 4*, as theory. Now, we will go through a hands-on journey using a real-world example. Before jumping into the example, let's brush up on the concept quickly in a few points:

- Define the list of secret keys in XRD using the `ConnectionSecretKeys` attribute.

- Define the namespace and secret name under the respective resource using the `WriteConnectionSecretToRef` configuration.

- Finally, populate the secret keys defined in the XRD using the `ConnectionDetails` configuration.

We will expand the hands-on example used for resource reference with nested XR to learn configurations for storing the secret. We created an S3 bucket, its policy, and an IAM user to access the bucket in that specific example. The example will not be fully finished until we extract the bucket details and IAM credentials into secrets. That is what we will exactly try to do in this example. The bucket details are already available in the Bucket resource, but we need to create a new resource named `AccessKey` attached to the created user for IAM credentials. The following diagram will represent the two XRs, their resources, and the secret key storage structure:

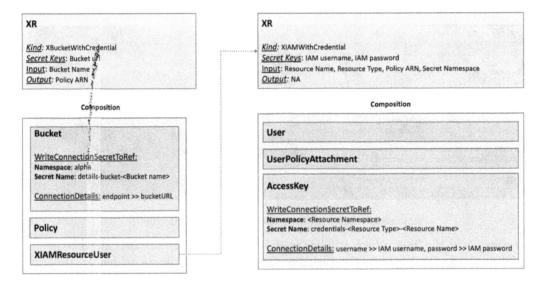

Figure 6.14 – Secret propagation

Let's look at some of the critical code snippets from the example. Following is the code from `XIAMWithCredential` and `XBucketWithCredential` to list the secret keys:

```
# List of secrets defined in XRD - XBucketWithCredential
connectionSecretKeys:
  - bucket_url


# List of secrets defined in XRD - XIAMWithCredential
connectionSecretKeys:
```

```
    - iam _ username
    - iam _ password
```

It was a simple list of secret keys under the `ConnectionSecretKeys` attribute in the XRD YAML. The secret name and storage namespace must be pushed to the resource to copy the secret key. Note that the namespace for the secret is automatically extracted out of the Claim. Following is the code from `AccessKey` and the `Bucket` resource to define the secret name and storage namespace:

```
# Secret name and namespace patching for Bucket resource
# Namespace to save the secret same as the resource namespace
- fromFieldPath: spec.claimRef.namespace
  toFieldPath: spec.writeConnectionSecretToRef.namespace
# Generate and patch the kubernete secret name
- fromFieldPath: spec.parameters.bucketName
  toFieldPath: spec.writeConnectionSecretToRef.name
  transforms:
  - type: string
    string:
      fmt: "details-bucket-%s"


# Secret name and namespace patching for AccessKey resource
# Namespace to save the secret is the same as the resource
- fromFieldPath: spec.parameters.secretNamespace
  toFieldPath: spec.writeConnectionSecretToRef.namespace
# Generate and patch the kubernete secret name
- type: CombineFromComposite
  toFieldPath: spec.writeConnectionSecretToRef.name
  combine:
    variables:
    - fromFieldPath: spec.parameters.resourceType
    - fromFieldPath: spec.parameters.resourceName
    strategy: string
    string:
      fmt: "credentials-%s-%s"
```

The final configuration we will look at is the actual copy of secrets into the keys defined at XRD. The following is the code from AccessKey and the Bucket resource to perform the same:

```
# Populate the connection secret keys from AccessKey secrets
connectionDetails:
- name: iam _ username
  fromConnectionSecretKey: username
- name: iam _ password
  fromConnectionSecretKey: password


# Copy the endpoint secret key to bucketURL for
connectionDetails:
- name: bucketURL
  fromConnectionSecretKey: endpoint
```

The example discussed is available at https://github.com/PacktPublishing/End-to-End-Automation-with-Kubernetes-and-Crossplane/tree/main/Chapter06/Hand-on-examples/secret-propagation. To fully experience the secret creation in the Kubernetes cluster, create the XR, composition, and Claim from the preceding link. The following screenshot covers the complete example execution:

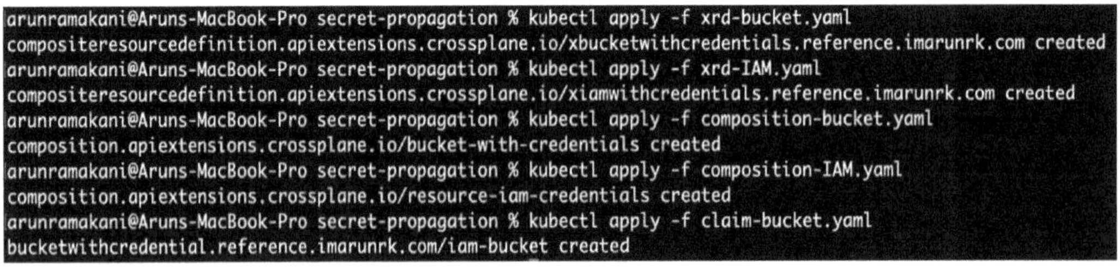

```
arunramakani@Aruns-MacBook-Pro secret-propagation % kubectl apply -f xrd-bucket.yaml
compositeresourcedefinition.apiextensions.crossplane.io/xbucketwithcredentials.reference.imarunrk.com created
arunramakani@Aruns-MacBook-Pro secret-propagation % kubectl apply -f xrd-IAM.yaml
compositeresourcedefinition.apiextensions.crossplane.io/xiamwithcredentials.reference.imarunrk.com created
arunramakani@Aruns-MacBook-Pro secret-propagation % kubectl apply -f composition-bucket.yaml
composition.apiextensions.crossplane.io/bucket-with-credentials created
arunramakani@Aruns-MacBook-Pro secret-propagation % kubectl apply -f composition-IAM.yaml
composition.apiextensions.crossplane.io/resource-iam-credentials created
arunramakani@Aruns-MacBook-Pro secret-propagation % kubectl apply -f claim-bucket.yaml
bucketwithcredential.reference.imarunrk.com/iam-bucket created
```

Figure 6.15 – Secret propagation

Once the resources are created in their entirety, you will see that the secrets are available inside the alpha namespace:

Figure 6.16 – Created secret

> **Information**
>
> May organizations prefer to store secrets in a key vault rather than Kubernetes secrets. There is an example on the Crossplane website to execute this integration at `https://crossplane.io/docs/v1.6/guides/vault-injection.html`. The Crossplane team is working on a more straightforward way to do this using an MR. The MR will represent the specific external vault resource and push the secrets accordingly. Keep watching the Crossplane release.

This concludes our exploration of secrets. The next section of this chapter will use the Crossplane Helm provider to deploy an application in a remote Kubernetes cluster. It will continue what we looked at in *Chapter 5*, in the *Managing external software resources* section.

Helm provider hands-on

It is exciting to introduce this aspect of Crossplane. It is precisely the crossroads where it unifies infrastructure automation and application automation. After creating an infrastructure resource, we would be interested in doing additional operations. For example, after deploying a Kubernetes cluster, we would be interested in setting up Prometheus or deploying an application in the remote Kubernetes cluster. Helm Crossplane provider can perform this operation.

Similarly, after provisioning a database, we will be interested in creating tables. SQL provider can perform these activities from Crossplane. The examples open a way to define all application dependencies in Crossplane and package them along with infrastructure. This section will go through a hands-on journey to experiment with Crossplane Helm provider. We will use GCP to create a Kubernetes cluster. It will fit well within the free tier limits. The following diagram represents how the Helm provider works inside the Crossplane ecosystem to manage application deployment in a remote Kubernetes cluster:

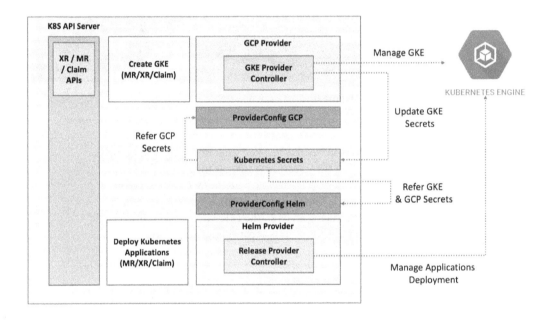

Figure 6.17 – Helm provider and GKE

Let's look at the details of how different components work together to manage applications using Helm in a few steps:

1. With your existing GCP provider and provider configuration, we can create a **Google Kubernetes Engine** (**GKE**) using the `Cluster.container.gcp.crossplane.io` MR.

2. Define the namespace and secret name in the MR to store the remote cluster credentials.

3. Install the Helm provider in the Crossplane control plane using the respective configuration YAML.

4. Next, set up the provider configuration for Helm provider using the Kubernetes credentials and GCP credentials stored in the Kubernetes secrets.

5. Now we can create the Helm releases in the remote GKE cluster using the `Release.helm.crossplane.io` MR.

Refer to the following code for the Helm provider configuration YAML:

```
apiVersion: pkg.crossplane.io/v1
kind: Provider
metadata:
  name: provider-helm
spec:
  package: crossplane/provider-helm:master
```

The following is the configuration for the Helm provider GKE. It requires credentials from both the Kubernetes cluster and the cloud provider. The secret reference under the credentials section refers to a specific Kubernetes cluster. The secret reference under the identity section refers to the GCP cloud credentials. The identity section of credentials may not be available for other cloud providers. Ensure that the Kubernetes APIs are enabled for the GCP cloud credentials:

```
apiVersion: helm.crossplane.io/v1beta1
kind: ProviderConfig
metadata:
  name: helm-provider
spec:
  # GKE credentials
  credentials:
    source: Secret
    secretRef:
      name: secret-gke-for-helm-deployment
      namespace: crossplane-system
      key: kubeconfig
  # GCP credentials
  identity:
    type: GoogleApplicationCredentials
    source: Secret
    secretRef:
      name: gcp-account
      namespace: crossplane-system
      key: service-account
```

Before applying the provider configuration, we must ensure that the GKE cluster is created and that its credentials are stored secretly. All examples of the Helm provider experiment are available at `https://github.com/PacktPublishing/End-to-End-Automation-with-Kubernetes-and-Crossplane/tree/main/Chapter06/Hand-on-examples/helm-provider`. Apply `GKE.yaml` to create the cluster. Refer to the following screenshot of GKE cluster creation, Helm provider installation, and provider configuration setup:

```
arunramakani@Aruns-MacBook-Pro helm-provider % kubectl apply -f GKE.yaml
cluster.container.gcp.crossplane.io/gke-for-helm-deployment created
arunramakani@Aruns-MacBook-Pro helm-provider % kubectl get cluster
NAME                         READY   SYNCED   STATE     ENDPOINT        LOCATION      AGE
gke-for-helm-deployment      True    True     RUNNING   35.223.47.160   us-central1   4m52s
arunramakani@Aruns-MacBook-Pro helm-provider % kubectl apply -f Helm-Provider.yaml
provider.pkg.crossplane.io/provider-helm created
arunramakani@Aruns-MacBook-Pro helm-provider % kubectl apply -f Provider-Config.yaml
providerconfig.helm.crossplane.io/helm-provider created
```

Figure 6.18 – GKE and Helm provider setup

Now we can start managing application deployment in the GKE cluster using Helm. The release is the MR construct available in Helm provider used to manage applications. Release MR has the following vital configurations:

- The `spec.forProvider.chart` configuration will hold basic information about the chart, such as the repository name, chart name, and version.

- `spec.forProvider.valuesFrom`, `spec.forProvider.values`, and `spec.forProvider.set` are the three different ways to provide the values for the Helm templated variables. If we set the values for the same variable in multiple ways, then the order of preference will be the same as the order mentioned previously.

- `spec.forProvider.patchesFrom` will be helpful in specifying post-rendering patches to override values at the last mile before deployment.

Refer to a simple `Release` configuration:

```
apiVersion: helm.crossplane.io/v1beta1
kind: Release
metadata:
  name: redis-crossplane-example
spec:
  providerConfigRef:
    name: helm-provider
```

```
forProvider:
  chart:
    name: hello
    repository: https://www.kleinloog.ch/hello-helm/
    version: 0.3.0
  namespace: default
```

Applying the preceding configuration will create the hello world example in the GKE cluster. Refer to the following screenshot with application deployment:

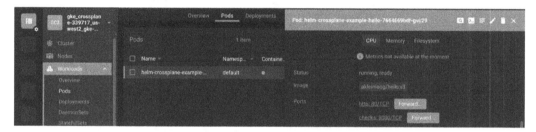

Figure 6.19 – Crossplane Helm deployment

The same `Release` MR from the Crossplane Helm provider also handles upgrades to our initial release of the Helm chart. We can change the required chart information, values, or patches and re-apply the YAML to upgrade our release. Apply `Helm-test-deploy-upgrade.yaml`, which changes the container version to move our release version. Before creating an upgraded release, the `Release controller` MR checks any actual change to the configuration. It will make sure that there are no unnecessary releases. Refer to the following screenshot showing an upgraded release:

Figure 6.20 – Crossplane Helm release upgrade

This concludes our Helm experimentation for now. The following section will rescue us from code and help us learn some guiding principles to define the XR API boundaries.

Tip

In all our examples, we referred to composition directly with its name in the Claim/XRs. We can also refer to the composition using label selectors after adding the respective labels in the composition metadata.

Defining API boundaries

We expect platform engineers to compose all the infrastructure and application automation concerns in XR APIs. How to define an API boundary is a bit tricky. It's because many conflicting trade-off points are influencing the API boundaries. Let's start with the fact that we wanted to compose every resource required for an application and its infrastructure in a single composition. Here are some considerations that will not allow us to do that:

- There would be many security and architecture policies that need to centralize. We cannot add them again and again in multiple compositions.

- Some resources may have compliance requirements and must be composed separately for audit purposes.

- Overly big compositions are difficult to read, understand, debug, and refactor.

- Testing a bulk composition is difficult.

- Every application will require some customization to the infrastructure recipe, provided we have a bulk composition. We will fork the main code for customization. It will be challenging to maintain as we grow.

- Specific infrastructure such as the network layer is owned by a particular team. It must be composed separately and referred to in the required XR.

There could be more reasons depending upon your organization's realities. In summary, we must build small XR APIs and organize them together with resource references and nested XRs. As soon as we talk about small XR APIs, API boundary questions arise. Which are the resources that make sense to be composed together? It is something like what we do in the world of microservices. A merge versus a split trade-off is something that we always do in microservices.

> **Tip**
> It's challenging to get the boundary correct on the first go. We should do our initial trade-off analysis, which provides an initial boundary and then evolves in iterations based on real-world experience.

Earlier in our S3 bucket example, we composed the bucket and its policy in a single XR. The second nested XR was holding the IAM user and policy attachment resource. This design can ensure that the IAM XR can be used with other resources.

> **Information**
>
> Later in *Chapter 10, Onboarding Applications with Crossplane*, we will do a trade-off analysis of a hands-on journey example to analyze the impact of different API boundaries.

The following diagram covers different factors influencing the trade-off analysis:

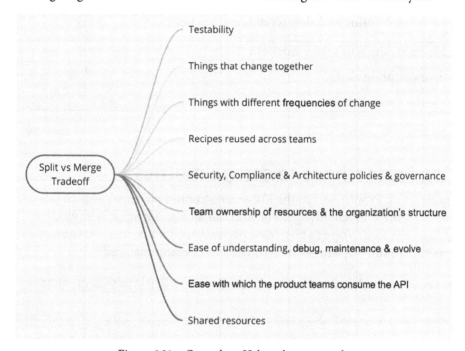

Figure 6.21 – Crossplane Helm release upgrade

This concludes the API boundary discussion. In the following section of the chapter, we will explore monitoring the Crossplane control plane.

Alerts and monitoring

Prometheus and Grafana, the popular tools in the Kubernetes world, can be used for Crossplane monitoring as well. Before starting, we should ensure that the Crossplane pod can emit metrics. It is as simple as setting the metrics parameter to true (--set metrics.enabled=true) during the Helm deployment of Crossplane. We can do it either at the first Crossplane release or upgrade the Helm release using the following command:

```
# Fresh install with metrics enables
helm install crossplane --namespace crossplane-system
```

```
crossplane-stable/crossplane --set args='{--debug}' --set metrics.
enabled=true
# Helm upgrade with metrics enables
helm upgrade crossplane --namespace crossplane-system
crossplane-stable/crossplane --set args='{--debug}' --set metrics.
enabled=true
```

We can split the monitoring and alert setup into three parts:

- Enable Prometheus to scrape metrics.
- Set up monitoring alerts.
- Enable the Grafana dashboard.

We can start first with metric scraping.

Enabling Prometheus to scrape metrics

First, we must set up Prometheus in the Kubernetes control plane. We will do this installation using the Prometheus operator. You can look at the quick start guide at `https://prometheus-operator.dev/docs/prologue/quick-start/`. The following are the simple steps to get the Prometheus operator installed:

```
# Step 1: Clone the Prometheus operator repository and switch to
the folder
git clone https://github.com/prometheus-operator/kube-prometheus.
git
cd kube-prometheus
# Step 2: Execute the initial setup instructions
kubectl create -f manifests/setup
# Step 3: Install the operator
kubectl create -f manifests/
# Step 4: Once the pods are up and running, view the dashboard
after the port forward
kubectl --namespace monitoring port-forward svc/Prometheus-k8s
9090
http://localhost:9090/
```

All configurations required for the monitoring example are available at `https://github.com/PacktPublishing/End-to-End-Automation-with-Kubernetes-and-Crossplane/tree/main/Chapter06/Hand-on-examples/monitoring`. Once you have Prometheus installed, the next step is to ask Prometheus to scrape the metrics from the Crossplane and GCP provider pods. We need to add `ControllerConfig` to the GCP provider to define the metrics port. Configuration of the same is available in `GCP-Provider.yaml`. Then, we can configure `PodMonitor`, which instructs Prometheus to scrape metrics from a specific POD at a given port. Configuration of the same is available in `monitor.yaml`. Once these steps are done, we can start looking at the controller reconciliation metrics in the Prometheus console. Create a GCP `CloudSQLInstance` instance with an incorrect configuration, which will not reconcile, and look at the reconciliation failure metrics. The following is the Prometheus query for the reconciliation failure metrics from `CloudSQLInstance`:

```
sum_over_time(controller_runtime_reconcile_errors_total{namespace="crossplane-system", controller="managed/cloudsqlinstance.database.gcp.crossplane.io"}[5m])
```

Refer to the following screenshot where we are looking at the reconciliation failure metrics for `CloudSQLInstance`:

Figure 6.22 – Metrics

The next step is to set up monitoring alerts for the reconciliation failure scenarios.

Setting up monitoring alerts

We can also set up alerts for this reconciliation error using the following configuration. It may be too much to trigger alerts for every reconciliation failure. Additionally, some of the reconciliation errors are expected scenarios. For example, if one resource is referring to another, the referring resource will fail to reconcile until the referred resource is provisioned. The following alert configuration is configured to throw an alert only if the reconciliation error exceeds 20 times within a 5-minute window:

```
apiVersion: monitoring.coreos.com/v1
kind: PrometheusRule
metadata:
```

```
name: sql-alerts
namespace: crossplane-system
labels:
    app.kubernetes.io/part-of: crossplane
spec:
  groups:
  - name: Crossplane
    rules:
    - alert: ReconciliationFailure
        expr: sum _ over _ time(controller _ runtime _ reconcile _
errors _ total{namespace="crossplane-system", controller="managed/
cloudsqlinstance.database.gcp.crossplane.io"}[5m]) > 20
        for: 5m
        labels:
          severity: page
        annotations:
summary: '{{ $labels.controller }} reconciliation has been
failing for more than 20 time in the last 5 minutes.'
```

Refer to the following screenshot where we are looking at the reconciliation failure alert:

Figure 6.23 – Alerts

Information

The list of metrics emitted by Crossplane is an area to be improved. We should get detailed metrics around compositions and Claims. We can expect more enhancements happening soon from the Crossplane community.

The final step involves setting up the Grafana dashboard to visualize the metrics and errors better.

Enabling the Grafana dashboard

Finally, we can set up a Grafana dashboard to visualize the metrics, logs, and alerts. The Grafana dashboard will already be installed in the cluster as part of the Prometheus Operator. What we must do additionally is to set up a dashboard for the Crossplane control plane. At `grafana.json` in the Git repository, I have added a sample dashboard configuration from the Crossplane community. Import the JSON into Grafana and look through the metrics. Refer to the following Grafana screenshot, which indicates that CloudSQLInstance is the active controller running the reconciliation loop:

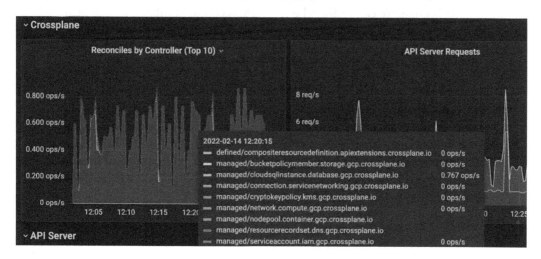

Figure 6.24 – Grafana metrics view

We will conclude the monitoring concepts here and move on to the final section of the chapter, which covers a few troubleshooting patterns.

More troubleshooting patterns

We explored different troubleshooting patterns in the last couple of chapters and earlier in this chapter. It covered ways to look at resource references in the resource description to move from composition to the MR, using the event logs in the resource description, and enabling Crossplane/provider pod logs to debug. This section will add a couple more debugging skills to enhance our platform development skills. The following are the new patterns we will look at:

- **Pause the Crossplane**: Sometimes, there may be a requirement to stop the controller reconciliation loop in order to debug resource issues. We can simply edit the Crossplane deployment to make the replication count to zero. This is the simplest way to pause Crossplane during our debug window. Once our debugging is done, we can restore the replication count. Similarly, we can also pause the Providers. We reduce the provider replication count to zero using `ControllerConfig` (the configuration we used earlier to enable debugging and expose the metrics port).

- **Hung resources**: Sometimes, we may notice that the resources are hung, and we cannot delete them. It should be because of the finalizers. We should patch the resource with an empty finalizer and delete it again. This will guarantee resource deletion only in the Crossplane control plane. It is not guaranteed that the resource is deleted from the external provider. We must visit the external provider console to validate whether the resources are deleted. The following code will render the finalizers empty:

```
kubectl patch <resource-type> <resource-name> -p
'{"metadata":{"finalizers": []}}' --type=merge
```

Summary

This chapter covered many new Crossplane patterns required to build the state-of-the-art resource composing control plane. I am explicitly mentioning the resource composing control plane instead of the infrastructure composing control plane because we no longer compose only external cloud provider resources. We experimented with resource referencing, secret propagation, and Helm deployment with the help of hands-on examples. We also looked at setting up monitoring with another hands-on example. In addition to these hands-on journeys, we also learned some debugging skills and ways to define our API boundaries.

The next chapter will cover different ways to extend and scale Crossplane.

7

Extending and Scaling Crossplane

The chapter will deep-dive into a few characteristics that make Crossplane extendable and scalable. The initial sections of the chapter will discuss developing new Crossplane providers for external resources that are not yet supported. We will examine the standards to be considered while designing a provider and the approaches available to make provider development comparatively comfortable. The following sections will cover configuration, the method to package the XR/Claim APIs, and how to test our XR. The final part of the chapter will cover different patterns supported by Crossplane to scale the control plane into a multi-tenant ecosystem.

The following are the topics covered in the chapter:

- Building a new provider
- XRD detailed
- A framework to build a provider
- Packaging and distribution of XR/Claim
- Testing the configurations
- Multi-tenant control plane patterns

Building a new provider

Crossplane providers are nothing but a bundle of related **Managed Resources** (**MRs**). MRs are opinionated custom controllers and custom resources combined into one-to-one mapping with external resources, enabling us to manage those resources from Kubernetes. Onboarding new resources as MRs into an existing or new provider is a time-consuming process. The Crossplane community has worked hard to onboard most of the essential resources with Crossplane-native controllers in the last few years. With the recent development in the Crossplane community to auto-generate, Crossplane providers from Terraform provider enabled 100% resource coverage for all cloud resources. In addition to the provider for all primary cloud providers, we also have providers for other external resources, such as GitLab, Helm, SQL, and Argo CD. Visit the Crossplane website or Upbound Registry to explore the available providers.

When we attempt to automate all application and infrastructure concerns, we might end up in a scenario where some external resources do not have a provider. You might decide to onboard a new provider yourself – for example, currently, there is no provider to manage Bitbucket repository resources. This section of the chapter will cover the essential aspects of the **Crossplane Resource Model** (**XRM**) required for new provider developers and explore the frameworks available to ease the development process. First, we will look at XRM, an opinionated subset of the **Kubernetes Resource Model** (**KRM**).

> **Tip**
> Looking at the standards defined by the XRM to support the new provider development also enables us to understand existing MRs much better. It will further help us build our XR with the same XRM standards. Following these standards with the XRs will enable a uniform and easy understanding for the consuming teams.

The following section will dive into the details of the XRM specification.

XRM detailed

Being an extension of the KRM, XRM inherits most of the standards. As discussed in *Chapters 3* and *4*, Crossplane inherits many standards from the Kubernetes CRDs. The opinionated XRM standards over the basic Kubernetes standards define a uniform bridge between Kubernetes and the external resource. The XRM standards cover the following characteristics:

- Configuration fidelity
- Spec and status configuration

- Naming the custom and external resources
- Configuration ownership
- Sensitive input and output fields

We can dive into the details of each of these characteristics in the following sections.

Configuration fidelity

The MR should have all possible fields available for configuration in the external resource API. It delivers every configuration combination at the control plane level. We should leave the abstractions to encode policy and the recipe creation for the platform developers via XR. Every field available for configuration in the external resource API should be present in the MR with the same name. Having the same name for the attributes will make it easy for users to compare and troubleshoot. Before starting a new provider's development, it's essential to ensure that the resources have well-defined CRUD APIs with granular controls. Control theory implementation cannot work well without such API standards. The following section will talk more about the API input and output.

Spec and status configuration

Spec is the configuration section that acts as the input of the API, and **status** is the attributes that are the output of the API. Let's look at the spec section first. We can have three types of configuration knobs in the spec section – initial initialization, late initialization, and immutable configuration are the three types. Initial initializations are the configurations used while resource provisioning as configured by the users. Many of the parameters fall under this category. The database version in RDS provisioning can be an example of initial initializations. The late-initialization attributes are initialized with default values by the provider when the resource is created and later updated with the actual value, configured by the user in the reconciliation loop. It is typically helpful for external resource fields that are only available after creating the resource. Resource tagging with labels is the most common example.

Immutable configurations are attributes whose value cannot be changed after initial provisioning – for instance, we cannot change the RDS region after creating the resource. Such configurations should be marked immutable in the MR. A reconciliation failure event is made if a user attempts to update the same. The mandatory fields required from Create and Update API operations should be marked as required fields in the MR. Optional fields should be of `pointer` type – a standard inherited from Kubernetes. We do this because some non-pointer types may resolve to zero when the user does not specify any value – for example, we should use `*bool` instead of `bool`. This concern is not applicable for a required field, as the value has to be specified by the user.

> **Information**
>
> Some fields can hold `struct` type value, as the underlying resource API supports it that way. We saw an example of this in the previous chapter. The `Policy` object from AWS IAM was taking a JSON policy as an input.

The *status* section represents the current state of the external resource. It holds attributes observed back from the external resource after every reconciliation loop. The fields that can be recreated if deleted are eligible candidates to be added to the *status* section. If a configuration and its sub-attribute represent another external resource, we should not include the relevant fields both in *spec* and *stats* – for example, we can also cocreate the subnet while creating Azure Virtual Network. In other words, the cloud API for Azure Virtual Network also has a section to define the subnet. From the Crossplane control pane, we should manage this as two unique resources with references.

Naming the custom and external resource

Each MR represents a unique kind, categorized under a specific API group, and it can exist under multiple versions. It is a behavior that is inherited from the Kubernetes standards. We have two important names when we create an object of a specific MR kind:

- The name of the resource within the control plane
- The name of the resource in the external ecosystem

The resource name in the control plane can also be called the **Custom Resource** (**CR**) name. At the end of the day, be it MR, Claim, or XR, it's an opinionated CR. When the user creates the MR/XR/Claim, they will provide a name parallel to the kind, version, and API group. This name will uniquely identify a resource within the cluster. If it's an MR/XR, the name is unique throughout the cluster. In the case of a Claim, it's unique per namespace. As these names are not autogenerated, the consuming teams may choose a random value. Such random names are challenging to cross-reference resources. A predictable naming strategy and labels will help resource references be straightforward. We can govern such standers in the configuration using admission controller tools such as **Open Policy Agent** (**OPA**).

The external name is the resource's name in the external ecosystem. The resource provider may or may not allow us to influence the name – for example, with the S3 bucket, we can determine the bucket's name in AWS. We control this name through properties such as ID, name, or UID. If the user does not configure these attributes, the controller can autogenerate with the `<CR Name>-<Namespace>-<Random 5 character assigned by Kubernetes>` template. Amazon VPC is a typical example where we cannot influence the name. AWS generates a unique ID called `vpcID` as a name. If we wish to link our MR/XR/Claim to an existing external resource, we can add an `crossplane.io/external-name` annotation with the resource's name to the external ecosystem. If a resource with a specified name does not exist, a new resource with the provided name is created. If it's a resource where naming is not allowed, the final name is copied back to the `crossplane.io/external-name` attribute. When designing our new MR, we need to keep these naming behaviors in mind.

Information

The KRM also recommends adding labels to the resources in the external environment with the kind, the CR name, and the provider name. It will help use cases such as identifying resources in the external environment, bulk operations based on the label, monitoring, and debugging. Note that all resources may not support such tags.

Configuration ownership

Some fields are relevant only to the control plane, both in the Spec and the Status section – for example, `ProviderConfigRef` in the Spec section indicates which credentials to use. We have other attributes that apply to the external resource only. There can be a conflict between these fields – for example, `CreationTimestamp` can exist at both the cluster and external resource levels. The KRM proposes the `spec.forProvider` and the `status.atProvider` sections as the owner for the external resource-related fields.

Sensitive input and output fields

There can be sensitive information in both the input and the output fields of an MR. Configuring the database master password is an example of sensitive data in the MR input. The IAM AccessKey credential is an example of sensitive data in the MR output. It is not sensible to directly expose these field values in the MR/XR/Claim. In the case of sensitive information in the MR output, the Kubernetes Secret should store the data. The Crossplane community is also coming up with *vault* integrations to publish credentials. Similarly, input fields with secret information should be confidential references. The controller should fetch the value for such fields from the secret source. The naming convention for such a field in the MR should be `<Field name>SecretRef`. The `MasterPasswordSecretRef` field in the `RDSInstance` MR is an example of such a field.

KRM is an area that evolves continuously with new requirements. Creating KRM standards to support resource reference in a Terraform-generated Crossplane provider is in progress. Another example of an in-progress standard is using Crossplane only as an observer, especially for existing resources managed by other automation tools such as Helm and Terraform. Regularly checking the Crossplane Slack channel or release notes can keep us updated on evolving standards. This concludes the KRM discussion. In the next section, we will look at a couple of approaches available for provider development.

Framework to build a provider

To bring down the cognitive load required to develop a new provider, the Crossplane community has identified a couple of comparatively painless ways, as follows:

- **Native provider development**: We have a template provider available at `https://github.com/crossplane/provider-template`. We can use this repository as a basic template for creating a new provider. The template has `ProviderConfig`, which can read credentials from the Kubernetes Secrets to manage external provider authentication and authorization. It also has a sample MR along with a controller. We can add all our new MRs and respective controllers with appropriate control theory implementation, using the CRUD APIs of the external resources. The video at `https://www.youtube.com/watch?v=dhuqH308Tc0` walks us through the provider development using provider template with a hands-on example.

- **Generate provider from Terraform providers**: Generating a native provider might be time-consuming. The community has developed a brilliant idea to auto-generate a Crossplane provider using the Terraform providers. The Terraform community has spent the last decade developing many production-ready provider modules. The Crossplane community has created a code generation pipeline called Terrajet that takes Terraform provider modules as input and converts them into a KRM-compliant Kubernetes controller. These controllers use the Terraform CLI to perform `Create`, `Read`, `Update`, and `Delete` operations on the external resource. The configuration knobs from the MR are converted to JSON and used as input to the `CRUD` operations, using the Terraform CLI. Step-by-step instruction on how to develop such a provider is available at `https://github.com/crossplane/terrajet/blob/main/docs/generating-a-provider.md`.

Most of us may not have a requirement to develop a provider ourselves. Already, we have 100% resource coverage for all significant cloud provider resources. The Crossplane community will create a Terrajet Crossplane provider soon for every available Terraform provider, provided we have a less cognitive load. If you still have a requirement, this section will have guided you in the appropriate direction. Additionally, learning the KRM will have taken your understanding of MR to the next level. The book's next section will cover ways to package and distribute the XRs/Claims.

Packaging and distribution of XR/Claim

Crossplane configuration is a way to package our XR and Claim APIs. This packaging will help us reliably establish these APIs into any Kubernetes cluster where Crossplane is enabled. Crossplane configuration is primarily a composition distribution mechanism. Along with the distribution, we also can manage versions and dependencies. We may end up using Crossplane configuration for three different use cases:

- It is useful when a large organization wants more than one control plane distributed across different team boundaries.

- It is also useful when someone is interested in building a control plane platform to sell as a product.

- Open source developers who want to share their XR/Claim recipes with the community can also use the Crossplane configuration.

This section of the chapter will go through a hands-on journey so that you can learn Crossplane configurations. To start with, let's look at packaging and distribution.

Packaging and distribution

With Crossplane configurations, we pack the compositions as an OCI-compliant image for distribution. Packing the compositions as configurations is a simple two three-step process:

1. As a first step, create a `crossplane.yaml` file. It's a simple YAML of `Configuration.meta.pkg.crossplane.io` kind. It defines the configuration name, the minimum-supported Crossplane version, and the dependencies. Both the provider and another configuration can be defined as dependencies. The following is the sample configuration YAML:

```
apiVersion: meta.pkg.crossplane.io/v1
kind: Configuration
metadata:
  name: aws-bucket
spec:
  crossplane:
    version: ">=v1.6.0"
  dependsOn:
    - provider: crossplane/provider-aws
      version: ">=v0.23.0"
```

This example is available at `https://github.com/PacktPublishing/End-to-End-Automation-with-Kubernetes-and-Crossplane/tree/main/Chapter07/Hand-on-examples/configuration`. We will pack our last Amazon S3 bucket example in a configuration named `aws-bucket`, which is designed to work in any Crossplane version greater than or equal to v1.6.0. The AWS Crossplane provider is added as a dependency for the package.

2. Next, we will execute the Crossplane CLI command build to generate the `configuration` package. We will run the command in the folder where we have the `crossplane.yaml` file and all our compositions. This step will output a file with a `**.xpkg` extension. Once the package is generated, the next step is to push the package into any OCI-compliant image registry. The Crossplane CLI push command will create an OCI-compliant image and move it into the registry. The following are the commands to build and push the configuration:

```
# Build the configuration package
kubectl crossplane build configuration
# Push the image into an image registry
kubectl crossplane push configuration arunramakani/
aws-bucket:v1.0.0
```

3. If you don't have the Crossplane CLI set up, use the following two commands:

```
# Download and run the install script
curl -sL https://raw.githubusercontent.com/crossplane/
crossplane/master/install.sh | sh
# Move the plugin to the bin folder
sudo mv kubectl-crossplane /usr/local/bin
```

Refer to the following screen for the installation in macOS:

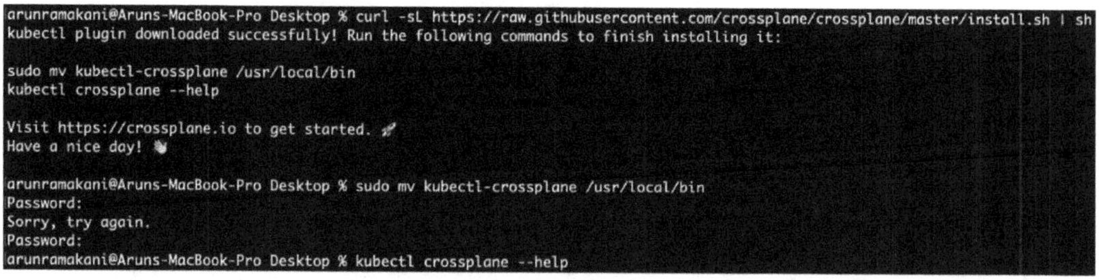

Figure 7.1 – The Crossplane CLI install

Note that the default image registry is the Docker hub. You should log in to the Docker hub in your CLI using `docker login`. You can see that the image is available at `https://hub.docker.com/repository/docker/arunramakani/aws-bucket`. You can also configure another image registry of your choice.

> **Information**
>
> Note that each composition and XRD combination is kept in different folders (`Bucket` and `IAM`). It will help the `build` command map and validate the composition and XRD combination. You will see a build error if all compositions and XRD are kept in the same folder.

Refer to the following screenshot, where we create the configuration and OCI-compliant image:

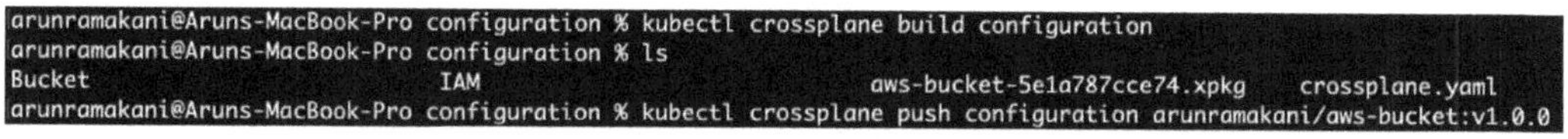

Figure 7.2 – Package and push the OCI configuration image

Once the configuration is available as an image in the registry, we can install the `configuration` package into the Crossplane control plane and start using the composition. We will look at that in detail in the following section.

Installing and using the configuration

We have two options to install the configuration in the Crossplane ecosystem. The first option is to use the following Crossplane CLI:

```
# To install a new configuration package
kubectl crossplane install configuration arunramakani/aws-bucket:v1.0.0 aws-bucket
# To upgrade an existing configuration package
kubectl crossplane update configuration aws-bucket v1.1.0
```

The CLI takes the name and image as two parameters for installation. It will even install the dependent AWS provider that we defined. Once you have the configuration installed, look at the details of the configuration and ConfigurationRevision with the following command. Every configuration update will create a new ConfigurationRevision, and only one revision will be active at a given time:

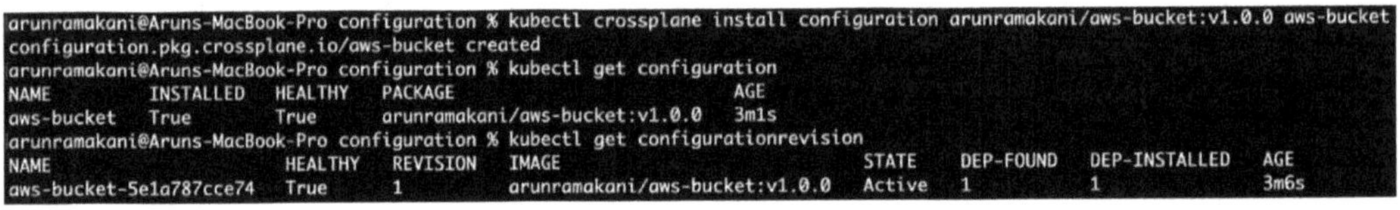

Figure 7.3 – Install configuration

The second way to install configuration is using the `Configuration.pkg.crossplane.io` kind YAML. Note that there are two configuration kinds in the Crossplane core with different API groups. The first API group, `meta.pkg.crossplane.io`, is used for building the `Configuration` package. The second API group, `pkg.crossplane.io`, is used for installing the `Configuration` package. The following is a sample configuration YAML:

```yaml
apiVersion: pkg.crossplane.io/v1
kind: Configuration
metadata:
  name: aws-bucket
spec:
  package: arunramakani/aws-bucket:v0.7.0
  packagePullPolicy: IfNotPresent
  revisionActivationPolicy: Manual
  revisionHistoryLimit: 3
```

We have not used the `PackagePullPolicy`, `RevisionActivationPolicy`, and `RevisionHistoryLimit` attributes in the CLI. `PackagePullPolicy` works very similar to `ImagePullPolicy` with other Kubernetes kinds. `RevisionActivationPolicy` can hold either `Automatic` or `Manual`, with `Automatic` as the default value. When it is automatic, the new XRs from the package are installed, XRs from the old package will become inactive, and the new XRs will become active to take charge of the resource reconciliation. You will see two ConfigurationRevisions after upgrading the `Configuration` package by one version increment. When a new ConfigurationRevision is created, you can also see that CompositionRevision is made for the changing composition. Refer to the following screenshot with one active and one inactive ConfigurationRevision:

```
arunramakani@Aruns-MacBook-Pro configuration % kubectl crossplane install configuration arunramakani/aws-bucket:v1.0.0 aws-bucket
configuration.pkg.crossplane.io/aws-bucket created
arunramakani@Aruns-MacBook-Pro configuration % kubectl get configuration
NAME          INSTALLED   HEALTHY   PACKAGE                              AGE
aws-bucket    True        True      arunramakani/aws-bucket:v1.0.0   6s
arunramakani@Aruns-MacBook-Pro configuration % kubectl get configurationrevision
NAME                      HEALTHY   REVISION   IMAGE                                STATE      DEP-FOUND   DEP-INSTALLED   AGE
aws-bucket-5e1a787cce74   True      1          arunramakani/aws-bucket:v1.0.0   Active     1           1               116s
arunramakani@Aruns-MacBook-Pro configuration % kubectl crossplane update configuration aws-bucket v1.1.0
configuration.pkg.crossplane.io/aws-bucket updated
arunramakani@Aruns-MacBook-Pro configuration % kubectl get configuration
NAME          INSTALLED   HEALTHY   PACKAGE                              AGE
aws-bucket    True        True      arunramakani/aws-bucket:v1.1.0   2m34s
arunramakani@Aruns-MacBook-Pro configuration % kubectl get configurationrevision
NAME                      HEALTHY   REVISION   IMAGE                                STATE      DEP-FOUND   DEP-INSTALLED   AGE
aws-bucket-5e1a787cce74   True      1          arunramakani/aws-bucket:v1.0.0   Inactive   1                           2m39s
aws-bucket-9cf65cbc813d   True      2          arunramakani/aws-bucket:v1.1.0   Active     1           1               15s
arunramakani@Aruns-MacBook-Pro configuration % kubectl get compositionrevision
NAME                             REVISION   CURRENT   AGE
bucket-with-credentials-gn8xw    1          False     3m25s
bucket-with-credentials-pthgr    2          True      60s
resource-iam-credentials-stqf7   1          True      3m25s
```

Figure 7.4 – Configuration update and revision

If we have set `RevisionActivationPolicy` in `Manual` mode, we must edit the revision manually to make it `Active`. The `RevisionHistoryLimit` field is the maximum number of revisions that Crossplane will keep track of. The following section will investigate ways to test Crossplane configuration.

Testing the configurations

The platform developers require a way to test the XRs and configurations they develop. Also, many teams might be interested in practicing test-driven development. This section of the book will explore **KUbernetes Test TooL** (**KUTTL**) as the test tool for practicing test-driven development and configuration-testing pipelines. KUTTL is a declarative test tool that tests for the best Kubernetes controller states and CRDs. The critical feature of KUTTL is writing declarative test cases against the CR. Being able to work well with CRs and CRDs, KUTTL can also work well with XR, XRDs, and Claim. First, we will look at the basic installation and setup required.

Installing KUTTL

The KUTTL CLI is an extension to kubectl. To install KUTTL, we will first install Krew. This is a kubectl plugin manager that helps discover, install, and update kubectl plugins. To install on a macOS/Linux operating system, run the following script:

```
(
  set -x; cd "$(mktemp -d)" &&
  OS="$(uname | tr '[:upper:]' '[:lower:]')" &&
  ARCH="$(uname -m | sed -e 's/x86_64/amd64/' -e 's/\(arm\)\
(64\)\?.*/\1\2/' -e 's/aarch64$/arm64/')" &&
  KREW="krew-${OS}_${ARCH}" &&
  curl -fsSLO "https://github.com/kubernetes-sigs/krew/releases/
latest/download/${KREW}.tar.gz" &&
  tar zxvf "${KREW}.tar.gz" &&
  ./"${KREW}" install krew
)
```

Set the path, making sure we can access Krew in the terminal, using the following command:

```
export PATH="${KREW_ROOT:-$HOME/.krew}/bin:$PATH"
```

The instruction to install Krew in other operating systems is different. To download and install Krew on different operating systems, refer to https://krew.sigs.k8s.io/docs/user-guide/setup/install/. Once you have Krew installed, the KUTTL set is a simple step – executing the following instruction:

```
kubectl krew install kuttl
```

After executing the preceding command, you will have KUTTL successfully installed in your local environment. The next step is to look at the anatomy of the KUTTL project and the basics of setting up tests.

KUTTL test setup

There are three critical components to the KUTTL setup. TestSuite.kuttl.dev is the first component and the core configuration. It holds the configuration for the entire test suite. The following is a sample TestSuite configuration:

```
apiVersion: kuttl.dev/v1beta1
kind: TestSuite
```

```
# Information on k8s cluster to use for testing
startKIND: false
kindContext: gke _ crossplane-339717 _ us-central1 _ autopilot-
cluster-1
skipClusterDelete: true
# Commands to be executed for any initial setup before testing
commands:
  - command: kubectl apply -f init/
# Directory where all our tests are kept
testDirs:
- ./bucketwithcredential
```

Note that the name of the TestSuite file should be kuttl-test.yaml to enable
the KUTTL CLI to look for the test configurations. StartKIND, KindContext, and
SkipClusterDelete are some of the configurations that defined the Kubernetes
cluster for testing. We use an existing Kubernetes cluster from GCP in the preceding
example. With KindContext, I have specified the name of the Kubernetes cluster
from kubeconfig. Instead, we can create a new kind Kubernetes cluster for testing
and destroy the same at the end of testing. It will be beneficial for test pipelines. Refer
to KUTTL documentation to understand full cluster configuration options. The
Commands section of the TestSuite file will enable us to run all initialization. We are
initializing ProviderConfig in our hands-on example. Look at the complete example
at https://github.com/PacktPublishing/End-to-End-Automation-
with-Kubernetes-and-Crossplane/tree/main/Chapter07/Hand-on-
examples/test-configuration.

Note that our ProviderConfig does not carry any actual AWS credentials.
ProviderConfig is available in the init folder in the Git repository. We initialize
ProviderConfig to ensure that the Crossplane will not complain about missing
configuration. Remember that scope of testing is to validate whether our XR and Claim
are correctly converted into an MR. In other words, we will be testing our composition
logic. If an MR will create the resource as expected, it is the scope of provider testing. If
we use a dynamic kind Kubernetes cluster, we should install Crossplane, AWS provider,
and the configuration as part of the initialization. It is out of scope for us, as we use an
existing cluster.

Tip

Provide the actual cloud credentials in ProviderConfig if you wish to do
end-to-end testing, or if your test case depends on some status field that we will
get back from the external resource.

The second key component is the test folders and test steps. KUTTL will scan for the test case from all the subfolders of the TestDirs folder specified in the TestSuite file. Each folder is a test case. In our case, we have two test folders. The bucket-failure folder holds a test case that will fail, and bucket-success contains a test case that will pass. Each test case can have multiple steps executed in a specified order. KUTTL can recognize the step number from the filename prefix. Note that the files inside the test case folder (in the Git example) have a numerical prefix.

The final section is to implement the individual test step. Each test step can have the configuration we apply in the Kubernetes cluster (XR/Claim) and the respective assert configurations (MR). The asserts do not have to define the whole MR but can have the fields we want to validate. We can define multiple assert configurations in a single assert step file. In the preceding example, we apply the bucket Claim and validate whether the composition patches the bucket name correctly (refer to the bucketwithcredential folder from the Git repository example). We can even assert and validate status fields in the MR, provided with a valid ProviderConfig. Execute the test from the root folder where we have the kuttl-test.yaml file with the following command:

```
kubectl kuttl test
```

Note that the bucket-failure folder assert configuration has a different bucket name different from the bucket Claim; hence, the test case will fail. Refer to the following screenshot, where one test case fails and the other succeeds:

```
--- FAIL: kuttl (50.66s)
    --- FAIL: kuttl/harness (0.00s)
        --- PASS: kuttl/harness/bucket-success (16.55s)
        --- FAIL: kuttl/harness/bucket-failure (32.14s)
FAIL
```

Figure 7.5 – The KUTTL test case

Information

We just covered the fundamental aspects of KUTTL. To explore the tool in more detail, visit https://kuttl.dev/docs/.

The following section will discuss the possibility of using KUTTL as a **Test-Driven Development (TDD)** tool for XR/Claim development.

TDD

TDD is a software development practice where developers develop test cases first from the requirements and then write code that passes the test cases. It is an iterative model of development where we start with failing test cases and slowly evolve code to pass. There are a lot of benefits to using TDD, including clean code and full test coverage. It's beyond the scope of this book to look at all the benefits. This section will focus on using TTD in XR development. The following figure represents the iterative TDD process for an XR/Claim:

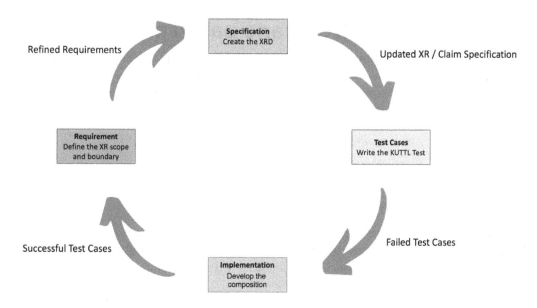

Figure 7.6 – TDD for XR/Claim

The following steps describe the stages in TDD for XR/Claim:

1. Define the API scope and boundary using the trade-off analysis discussed in the previous chapter.

2. Create the XR/Claim API specification using the XRD configuration, using the scope and boundary definition. It will be a requirement from the perspective of API consumers.

3. Define the final MR state of each resource in the composition. This only has an API implementation requirement. Organization policy and compliance requirements will become part of the implementation requirement.

4. Develop the KUTTL test cases for all the requirements defined in *Steps 2* and *3*. Run the test cases to see that all instances fail.

5. Develop the compositions and rerun the test cases. Iterate until all the test cases succeed.

6. Refine the API scope and boundary requirements to continue the cycle.

We can combine KUTTL with other Kubernetes ecosystem tools such as Skaffold to make our TDD easy. This concludes our discussion on testing the configuration/composition. In the final section of the chapter, we will cover the different ways to scale Crossplane into a multi-tenant ecosystem.

Multi-tenant control plane patterns

Typically, multiple product teams will have to access the control plane platform to take advantage of the composition recipes built by the platform engineers. The section covers different patterns that Crossplane supports to enable a multi-tenant control plane. The following are the key two patterns Crossplane users can choose from:

- Multi-tenancy with a single cluster
- Multi-tenancy with multiple clusters

Multi-tenancy with a single cluster

Multi-tenancy with a single cluster is a pattern where all the product teams use a single Crossplane control plane. The control plane is configured to enable multi-tenancy in the same cluster itself. The following facts describe what this setup will look like:

- The product teams are isolated with the namespace Kubernetes construct. Each product team should be assigned a namespace.

- As mentioned previously in an earlier discussion about the difference between XR and Claim, Claims are namespace-scoped and the XR is of cluster scope.

- Also, we discussed earlier that XRs are meant to be used only by the platform team.

- Precise RBAC can be applied to the Claim API in the given namespace, based on team member roles and the API group.

- An organization can implement RBAC, either using the default Kubernetes RBAC API or a general-purpose policy engine, such as the OPA (Gatekeeper).

In addition to the preceding points, each team should have different external provider credentials to track the usage, cost, monitoring, audit, and so on. It can be easily achieved using the ProviderConfig named after the namespaces. We should also be patching the ProviderConfigRef from the Claims namespace reference in the composition. The following figure shows the architecture that can help you visualize the single cluster multi-tenancy:

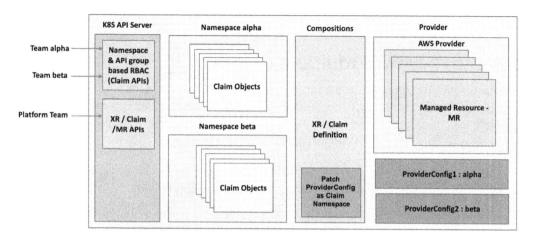

Figure 7.7 – The single cluster multi-tenancy

The following is the code snippet referring to the pattern:

```
# ProviderConfig for each team
# Match the ProviderConfig name to the namespace
# Example ProviderConfig for team-alpha
apiVersion: aws.crossplane.io/v1beta1
kind: ProviderConfig
metadata:
  name: team-alpha
spec:
  credentials:
    source: Secret
    secretRef:
      namespace: crossplane-system
      name: team-alpha-creds
      key: creds
```

Under each resource in the composition, patch `ProviderConfigRef` with the Claims namespace dynamically:

```
# Patch claims namespace to the ProviderConfigRef of the MR
patches:
  - fromFieldPath: spec.claimRef.namespace
    toFieldPath: spec.providerConfigRef.name
```

The following section will look at the multi-tenancy setup with multiple clusters.

Multi-tenancy with multiple clusters

Some organization setups may have multiple independent business units with different infrastructure requirements, such as monitoring, cost management, and compliance. We might have the condition to set up multiple Crossplane control planes. The following two basic patterns in Crossplane will support such an environment:

- **Configuration**: We can leverage the XR/Claim packaging mechanism discussed earlier in this chapter to develop and distribute XR/Claim reliably.

- **Nested crossplane**: One central Crossplane control plane can manage other Kubernetes clusters for each business unit. We can use the Helm provider from the central Crossplane cluster to set up Crossplane across the other clusters.

We can also attempt multiple Crossplane setups within a single cluster, one for every tenant/team. We can try this with tools such as `vcluster` or similar tools. It is an advanced pattern. What we attempt here is Kubernetes inside another Kubernetes. If you have a similar use case, try the setup using vcluster. This concludes the multi-tenancy discussion.

Summary

This chapter discussed various aspects that extend and scale Crossplane. We have taken a step-by-step journey to learn Crossplane from its basics to many advanced concepts in the last few chapters. We covered many nuances of building a state-of-the-art control plane for automation using Kubernetes and Crossplane. This takes us to the end of part 2 of the book.

The book's third part will explore an approach to managing configuration along with some configuration management tools and recipes. It will be a journey to unify application and infrastructure automation with Crossplane and other configuration management tools.

Part 3: Configuration Management Tools and Recipes

There is an overwhelming list of tools in the Kubernetes configuration management ecosystem where we don't have clear guidance on the trade-offs. This part of the book will guide cloud-native practitioners in choosing the tools for Kubernetes configuration automation that best suit the use case.

This part comprises the following chapters:

- *Chapter 8, Knowing the Trade-Offs*
- *Chapter 9, Using Helm, Kustomize, and KubeVela*
- *Chapter 10, Onboarding Applications with Crossplane*
- *Chapter 11, Driving the Platform Adoption*

8
Knowing the Trade-offs

In the previous few chapters, we learned a lot about Crossplane, from its basics to many advanced patterns. Also, we introduced the idea of a unified approach to both application and infrastructure automation. This chapter will step back to analyze and approach configuration management for unified automation holistically. The chapter is heavily influenced by the white paper *Declarative application management in Kubernetes* (https://goo.gl/T66ZcD) by Brian Grant, a Kubernetes Steering Committee emeritus. The white paper mainly covers the **Kubernetes Resource Model** (**KRM**). We will cover the topics in this chapter from the KRM and **Crossplane Resource Model** (**XRM**) perspectives. The chapter will define the scope of the unified automation approach. It will continue by looking into many more concepts, such as the tools available, common pitfalls along the way, and the trade-off in using different patterns. In a way, it's a revisit of the API boundaries discussion from *Chapter 6*, in more detail.

The following are the topics covered in the chapter:

- Unified automation scope

- Complexity clock, requirements, and patterns

- Open Application Model

- Specialized and extendable abstraction

- Impact of change frequency

Unified automation scope

Most of us will have different perceptions of what we mean by unified application and infrastructure automation. This section will help us understand the scope with a bit more clarity. Any application/product that we deploy is mostly a combination of custom-developed bespoke applications and **common off-the-shelf** (**COTS**) components. The term bespoke application means custom-written software for our requirements with a specific purpose. Bespoke applications are generally stateless, containerized workloads. From the perspective of Kubernetes, they are workloads that run on Pods, the basic unit of computing in Kubernetes. COTS components are generally stateful infrastructure components, such as databases, cache systems, storage systems, and messaging systems. In the cloud-native era, most COTS components are black-box, fully managed **software as a service** (**SaaS**) or **platform as a service** (**PaaS**). COTS vendors expose CRUD APIs to work with the resource with precise configuration knobs for security, scaling, monitoring, and updating concerns. These specific configuration knobs will support different use cases from consumers. When we talk about unified application and infrastructure automation, it's an approach to use the same set of tools, processes, and patterns to configure both bespoke applications and COTS deployment. In other words, bespoke applications can follow KRM standards, and COTS dependencies can comply with XRM standards, an extension of KRM. There are many advantages to such unification:

- In the modern software engineering discipline, the product teams are vertically sliced to bring maximum delivery velocity. These vertically sliced teams own both bespoke applications and their dependent COTS components. Unified tooling, processes, and patterns will significantly reduce the cognitive load.

- It can bring down the need for specialized teams to manage COTS components, accelerating the delivery velocity further.

- The configuration data for policies in bespoke applications and their COTS dependencies can be quite simple. We can validate policies such as architecture fitness functions, security concerns, and compliance standards in a single place and format.

- All COTS vendors can offer KRM-compliant APIs (MRs) as a universal application dependency and integration standard. It is already happening with Crossplane providers for all primary cloud resources. The list is growing to cover other external resources, such as Git Lab, Argo CD, Rook, and Cloudflare, to cover end-to-end automation.

The following section will cover a few requirements, patterns, and tools for approaching unified automation.

Complexity clock, requirements, and patterns

The configuration complexity clock is a concept that explains how configuration management can become complex over time. It explores different stages in the evolution of configuration management, its use cases, and its pitfalls. It was initially discussed from the perspective of application configuration in the blog post found here: `http://mikehadlow.blogspot.com/2012/05/configuration-complexity-clock.html`. We will look at the same concept from the Kubernetes configuration management perspective.

The configuration complexity clock

Let's say we are deploying our first application workload into Kubernetes. To start with, every Kubernetes artifact, such as Deployment, Service, and Ingress, can be managed as individual configuration artifacts. It may not be an issue when we operate on a tiny scale. But soon we will realize that there is an issue with consistently performing releases and rollbacks as the application configuration is spread over multiple Kubernetes resources. We will start thinking about packaging and release management tools such as Helm. Helm is a template-based configuration management tool. We will parameterize values with variables to enable configuration abstraction and expose only limited attributes. A typical example is setting the replica count for every deployment target (production and staging). Soon, the organization will realize the power of Kubernetes and decide to deploy further similar workloads in the Kubernetes ecosystem. As the configuration data for the new application looks identical to the initial Helm template, we will choose to parametrize more variables to support multiple applications with the same Helm template. It will be an excellent way to push the reuse agenda and minimize the configuration to maintain. A typical example of such new parameters would be a namespace, labels, and an application name.

Again, the organization will realize more benefits from Kubernetes and decide to experiment with a few new workloads. These workloads may have similar configuration data to the initial Helm template with minor customization. We would decide to fork the main Helm template to do the required customization. Template-based abstractions are complex to reuse when customization is required. After a certain length of time of the configuration clock running, we will see too many forks that are difficult to keep in sync. On the other side, many individual templates would have introduced new parameters to support new local use cases. This is a leak in the abstraction we created with the Helm templates. We would have parameterized all values and completely eroded abstraction as the clock ticks further. Have a look at this WordPress chart as an example: `https://github.com/helm/charts/tree/master/stable/wordpress`. The number of parameters has increased slowly into multiple folds from the initial commit. Entirely eroded templates are complex to read, and end users will find it challenging to understand the usage of each parameter. Users will find it challenging to see parameters that are not their concern. For example, developers may not know how to define Ingress parameters in the preceding WordPress Helm chart.

As the configuration clock ticks further, we must harden the configurations for security. Infrastructure operators will want to own the configuration as the application owners do not know how to configure the security parameter. The Helm post rendering feature can rescue us from the situation and help inject the security configuration as late binding. But it will still be challenging for the infrastructure operators to manage the post-rendering step in the deployment pipeline because of too many customization forks and unexpected rendering outputs. Some developers may decide to use a DSL such as Terraform to manage configuration as it is simple to read. DSLs inherit the same parameterization problem in input variables, and the post-rendering customization step is also challenging with DSL. Additionally, DSLs have the issue of a complex learning curve for developers and limited tooling support for concerns such as testing. Vulnerability or compliance scanning is another area where we may face tooling issues. The Crossplane configuration can also face similar problems as we scale over time. The idea of discussing these limitations is not to discourage any specific tool usage. Each tool has its use cases and trade-offs. For example, Helm packing works well when operating on a small scale and a few other use cases, such as release management. In a way, we will be looking at how we can use a combination of tools and patterns to overcome this limitation.

Configuration management requirements

To manage configurations without being trapped by the issues discussed regarding the complexity clock, we should keep a few guiding principles in mind. These guiding principles will be the technical requirements to perform trade-off analysis when selecting tools and patterns for configuration management. Baking these guiding principles into our configuration management recipes would allow us to evolve and scale with ease:

- Keep the configuration data readable for humans and machines to mutate along the configuration pipeline. Requirements such as environment-specific patching, security configuration patches, and policy as the configuration are best implemented as an automated step in the pipeline. Avoid maintaining configuration as code as they are challenging to manipulate. Rendered output from code may not always be straightforward for the machines to read.

- Configuration scanning requirements, such as audit, compliance, and security, work well with clean configuration as data. Keeping the configuration as data throughout the pipeline as much as possible and separating code that manipulates the configuration is vital to meet the evolving requirements. Many tools evolved around the Kubernetes ecosystem because configuration data is kept separate (in etcd) from code (controller) and has a standard data format with KRM.

- Segregation of concerns is another critical aspect of configuration management. Not all configuration knobs required to automate are meant for a single persona to define. We should manage the configuration so that different people can collaborate with ease.

- Build specialized abstractions to support customization with the reuse of configuration. For example, Pod is a fundamental configuration abstract over which Deployment, ReplicationController, DaemonSet, and so on are built. Later in the chapter, we will see more solid examples from the Crossplane and complete application management perspective.

- Version control the source configuration and the final mutated configuration, representing the desired state. Avoid modification to the configuration post applying to the cluster. Tools such as mutating admission controllers should be avoided. Late configuration binding may fail in ways that are hard to predict (still, the admission controller is suitable for policy validation).

- Use composing to bind the application and its dependent resource into a single API while maintaining separation of concerns with nested sub-APIs. Facilitate release management concerns such as discovery, life cycle management, and dependency management for the entire application bundle.

We will deep dive into all these guiding principles in the upcoming sections. The following section will examine different patterns available for configuration management and the trade-offs with each pattern.

Patterns and trade-off

Reusing the configuration as we scale is challenging. Most of the reuse patterns were discussed in the *The configuration complexity clock* section. The section will cover these patterns in more detail with their advantages and disadvantages. The following are the well-known approaches for configuration reuse:

- **Fork**: This is one of the frequently used methods to reuse configuration, primarily because of the ease of customization. Rebase is the strategy used to sync the configuration as things evolve. The rebasing process cannot be automated as humans must address merge conflicts. The method offers a high level of flexibility and comfort of maintenance in the short term. As time passes by, a couple of challenges arise. The forks will diverge a lot with time, creating challenges with rebasing. With independent agile teams managing different forks, the advantages of reuse can be overlooked to keep up with the evolving speed. I had a similar experience a year back when I decided to fork a code repository for customized deployment requirements. Different agile teams owned both forks. As the clock ticked further, code merge activity was ignored entirely, yielding to the delivery pressure. Finally, we ended up in a state where the two forks could never sync with thousands of conflicts. The fork solution works best when you quickly try a proof of concept. I will not recommend using them beyond that.

- **Template parameterization**: We discussed template-based parameterization in *Chapter 2*, and earlier in this chapter. It works well on a small scale but suffers leaking abstraction and team collaboration issues as we scale. Additionally, it will also push us toward using the fork pattern by making customization complex. Tools such as Helm fall under this category. As a template-based parameterization tool, Helm is extremely popular because of its other capabilities, such as application discovery and release management.

- **Patch/overlay**: This method would have the base configuration as pure data points and use the patch file to overlay the required variables for customization. We can replace values for an existing configuration knob or add a new configuration to the overall base. We can also use the technique as a post-rendering step in template-based abstraction tools such as Helm. The leaks are avoided as additional parameterization required for customization can be managed as a post-rendering patch. It's a popular method that is being used quite a bit at the moment, especially since the Kustomize tool came into existence. These tools do not require human intervention and can be automated as a step in the deployment pipeline.

- **Composition**: It is a technique where we compose the dependencies to build a higher-level API and multiple sub-APIs consumed by different persona. We have discussed this pattern well enough as Crossplane itself is a composing tool. Pull and push are two sub-patterns within the composition. In the pull model, the dependencies are directly referred to. Nested XR is a typical example from the Crossplane world. With the push model, the dependencies are referred across API indirectly through runtime binding. The push pattern is suitable for separating concerns and building extendable abstractions. Resource references between two independent XRs using labels is a push composition example. Composition and patch/overlay work well together to create customization. For instance, in Crossplane, we can use a patch to generate environment-specific composition (production/staging).

Composition, patch/overlay, and template-based tools can complement each other and combining them can help you to build a robust deployment pipeline. We will look at a few such hands-on examples in the following two chapters. The next section of this chapter will look at the **Open Application Model (OAM)**, an open source standard to define application and team-centric configuration management.

Open Application Model

OAM is an open source standard developed by Microsoft and Alibaba to guide us in creating higher-level abstraction APIs for application deployment. In other words, it's about creating a model for application configuration management using composition as a pattern. The standard focus on addressing the following three problems:

- **Developer focus**: Exposing developers directly to Kubernetes configuration management to deploy the applications will make them spend time figuring out infrastructure details rather than application focus. Hence, OAM attempts to keep developers focused on the application.

- **Vendor dependency**: Configuring applications usually tends to depend on the underlying infrastructure. Completely decoupling the application configuration from the underlying infrastructure can enable the portability of workloads. Kubernetes does this to an extent, but the area requires more work with cross-cutting concerns and COTS dependencies.

- **Extendibility**: Configuration management at scale can have many problems, especially while balancing reuse and customization. OAM proposed a model for reuse and customization for scale.

OAM proposes layered configuration management based on personas and composing abstractions accordingly to address the problems. OAM defines three personas to manage different application deployment concerns, as represented in the following figure:

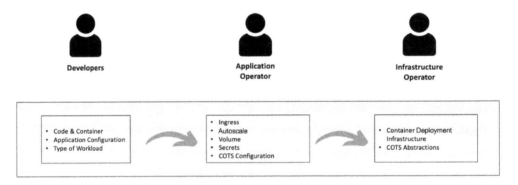

Figure 8.1 – OAM persona

KubeVela, the OAM implementation

The OAM community has developed KubeVela, a project that implements the specification. It is a CNCF sandbox project and a composing tool like Crossplane. KubeVela concentrates only on composing bespoke application configurations. But Crossplane composes both bespoke applications and COTS infrastructure/external dependencies. Both KubeVela and Crossplane can complement each other using the following two different patterns:

- **KubeVela composes Crossplane:** We can use KubeVela to compose bespoke applications deployment abstractions, and it can rely on Crossplane as the COTS external dependency provider. This pattern requires the Crossplane control plane to be present in the application workload cluster.

- **Crossplane composes KubeVela**: We can use Crossplane to compose abstractions for both bespoke applications and COTS infrastructure/external dependencies. For bespoke applications, it can use KubeVela through Crossplane Provider for Kubernetes. This pattern can work with a centralized Crossplane control plane or a distributed Crossplane control plane.

The following figure represents the KubeVela composes Crossplane pattern where both the Crossplane control plane and application workload cluster are the same:

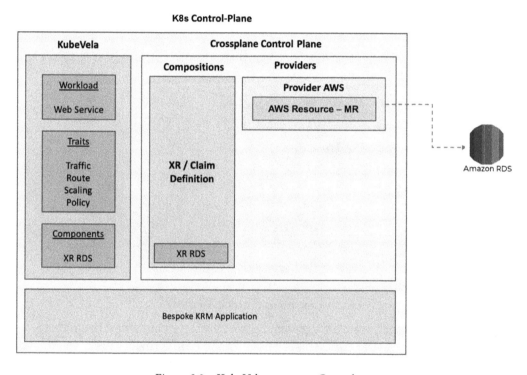

Figure 8.2 – KubeVela composes Crossplane

The preceding figure represents a web service workload with an RDS database as a COTS dependency. The workload construct of KubeVela defines the application workload type. Traits define all workload characteristics, such as route and traffic. Components help define the COTS dependencies. First, we will compose RDS into an XR API using Crossplane. Later, we can compose the XR as a component inside KubeVela. The following figure represents the second pattern where Crossplane composes KubeVela:

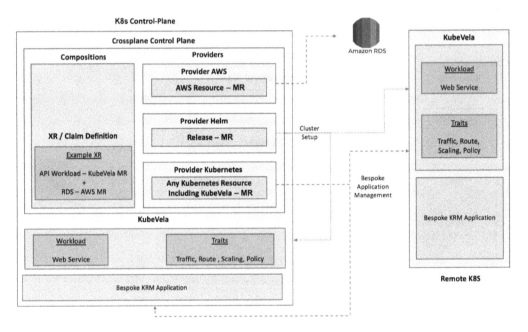

Figure 8.3 – Crossplane composes KubeVela

The figure represents application management both in the same cluster and in a remote cluster. Additionally, the figure represents the usage of the Helm provider to set up KubeVela and other cluster dependencies. We will cover a hands-on example of KubeVela in *Chapter 9*.

> **Information**
>
> OAM and KubeVela are exciting topics to explore further. Visit `https://oam.dev/` for more information.

The following section will cover two exciting patterns that can help platform teams to build extendable and reusable platform APIs.

Specialized and extendable abstraction

As we scale the number of applications deployed in Kubernetes, there could be an exponential proliferation of configurations to manage. Managing a high volume of KRM/XRM configuration files is prone to human error, is challenging to keep in sync, and requires a highly skilled workforce. Reuse is the key to keeping the volume of configurations low. But customization requirements at the individual team level will not allow us to reuse unless we have an easy and quick way to do so. Also, agile and product engineering practices will add additional pressure from a minimal external dependency perspective. Specialized and extendable abstraction is vital to address these problems. Let's start with specialized abstraction in the following section.

Specialized abstraction

Specialized abstraction is a technique where we build a basic abstraction and reuse the base abstraction to make specialized abstractions that handle custom requirements. It is nothing but the inheritance pattern from object-oriented programming. Also, this is a well-known pattern in core Kubernetes. Think of the Pod resource type as an example. Deployment, DemonSet, and Job are specialized abstractions that use Pod as their base. Let's look at some of the specialized abstractions we could build as a platform team:

- **Policy abstraction layer**: We may have policies per cloud resource as an organization. These policies could be from diverse areas, such as security, architecture, and compliance. If this is how the cloud strategy works in your organization, consider creating an abstract layer above the Crossplane provider with an XR for each resource. The layer can act as a solid base for creating recipes.

- **Database recipes**: Different product teams will have different database requirements. Some product teams will demand a geographically distributed database, and others may be happy with availability zone fault tolerance in a single region. A third team may require a PCI-DSS-compliant database to store payment transactions. We could create a base database XR and create a specialized XR above the base to fit each need.

- **Workload recipes**: Not every workload is the same. Web applications, RESTful APIs, scheduled jobs, and data transformation are some examples of workloads, and each will have different dependency requirements. For example, a scheduled job workload does not require Ingress. We create a primary workload with all common cross-cutting concerns and develop a specialized abstraction over the primary workload. We can do this with Crossplane or KubeVela or a mix of both.

These are just some general examples that may apply to your environment. Finding the correct abstractions that suit your environment will require a closer look at the organization structure, technical maturity, governance model, workloads, and so on. The base and specialized abstractions are typically developed by the platform team to be consumed by the application operator and developers. We may have a rare scenario where the application operator is building a specialized abstraction over the base developed by the platform team. We can look at extendable abstraction in the following section.

Extendable abstraction

Extendable abstraction is a technique where we build a partial abstraction and allow other teams to implement it fully. It is nothing but the polymorphism pattern from object-oriented programming. Also, this is a well-known pattern in core Kubernetes. Ingress is an example from core Kubernetes. Each managed Kubernetes provider will come up with its implementation for Ingress. Extendable abstraction is meant for both platform teams and application operators. Let's look at some examples of the pattern's usage:

- **Shared resources**: Let's consider VPC as an example of a shared resource. Multiple variants of the resources sometimes need to be created and used for different scenarios. We can have a standard label name/value strategy for such resources, and other XR recipes can choose one VPC with appropriate label references. This is polymorphic behavior and provides extendibility through dependency injection.

- **Nested XR**: The platform teams can create an XR recipe with dependencies missed identified. The missing dependencies can be implemented separately. Both the pieces can be composed with a nested XR pattern. Choose this pattern when the requirement for that missed dependency changes with every product team. An application recipe that leaves the database choice open is an example of nested XR behaving in a polymorphic way. The application operator can complete the recipe based on the specific product team's requirements.

Again, these are indicative examples. This concludes the abstraction discussion. In the following section of the chapter, we will look at configuration change frequency in detail.

Impact of change frequency

Looking at the configuration knobs from the perspective of frequency of change will help us categorize them between personas defined in the OAM. The change frequency detailing will bring in the discussion of ownership as these perspectives are linked. This section covers change frequency from the perspective of bespoke applications (KRM compliant) and COTS external dependencies (XRM compliant). We can classify configuration change frequency into the following three categories:

- **Change very often**: Generally, application KRM configuration parameters, such as image tags, release version, environment variables, secrets reference, and application configuration, change very frequently. These configurations are application-centric and owned primarily by the application developers. If we use a template- or DSL-based abstraction, they are good candidates to be exposed as variables. Suppose a plain configuration YAML is used instead of a template- or DSL-based abstraction, developers can own a version-controlled patch/overlay file. When composing is used as the solution, these configurations should be exposed to the developer with a high-level API.

- **Less frequent change**: Configurations such as application context-related information (namespaces, labels, references, and so on), resource constraints (memory, volume, and so on), and release-related knobs (replica count and rollout method) are examples of less frequently changing configurations. The preceding-mentioned configurations mainly vary per environment or change when there is a new operational requirement. For example, the number of replicas, namespace, or labels can change based on the deployment environment (production, staging, and so on). Irrespective of using template-based or DSL-based, or a plain YAML or composing, it's best to use patching as a mechanism to manage less frequently changing configuration values and choose the patch file based on the target environment in the pipeline. Patching is a choice because application operators own these configurations independently. Exposing them to developers will increase the cognitive load.

- **Rarely changes**: Rarely customized configurations are the basic structure of the core application. Platform developers should own these configurations. The basic structure is the core abstractions to minimize the cognitive load for application operators and developers. Also, it will summarize the policy requirements inside the recipe. Generally, we use overlay, specialized, and extendable abstractions to achieve multiple variants of core configurations required by the different product teams and workloads.

XRM change frequency

The XRM-compliant COTS external infrastructure dependencies mostly do not change frequently. For example, changes to the external infrastructure, such as region scaling, autoscaling setup, tuning security configuration, upgrade, and migration, will happen at a low phase after the initial setup. Building these infrastructure abstractions with XRs should be owned by platform teams. The application operator could do a few XR composing exercises to make a new, specialized recipe. In other words, they could create workload-specific abstraction using the nested XR pattern. As discussed earlier, XRM goes beyond infrastructure dependencies to manage the external application using providers such as Helm and SQL. These configurations could change frequently. For example, every application release could change the SQL schema of the database. Hence, the application operator persona can extend the existing recipes to meet the product team requirement.

Summary

This chapter discussed various aspects of unified configuration management of bespoke applications and COTS components at scale. We covered concepts such as different tools available for configuration management, common pitfalls along the way when we scale, and the trade-off in using different patterns. Also, we discussed how different combinations of tools and practices could complement each other under different circumstances.

The following two chapters will look at hands-on examples to try out a few recipes discussed in this chapter. The recipes will include KRM and XRM configuration management as we move toward end-to-end automation of the whole application and its COTS dependencies.

9
Using Helm, Kustomize, and KubeVela

This chapter concentrates mainly on configuration management for bespoke applications. The Kubernetes configuration for bespoke applications includes deployment, service, ingress, secret, configmaps, tags needed for governance, cross-cutting concerns, application security context, and other dependencies. Managing these configurations requires carefully choosing patterns and tools that fit the use case. Also, we must keep reuse, team collaboration, and scalability in mind. In the previous chapter, we lightly touched on a few tools such as Helm, Kustomize, and KubeVela for application configuration management. This chapter will be an opportunity to explore these tools in more detail.

The following topics are covered in the chapter:

- Application configuration management capabilities
- Using Helm for application deployment
- Hands-on chart development
- Customizing configurations with Kustomize
- Deploying application workloads with KubeVela

Application configuration management capabilities

Operating an application successfully in the Kubernetes environment requires a few capabilities from the perspective of configuration management. Here is a list of critical configuration management capabilities:

- **Packaging**: As discussed several times, deploying an application into Kubernetes involves configuring multiple resources. It requires a capability where we can package all these resources into a single bundle.

- **Life cycle management**: An application and all its dependencies must be executed into the cluster as a single deployment supporting the required release management constructs such as rollout, rollback, version management, and blue-green deployment.

- **Application discovery**: This is a capability that is required for day-to-day operations. It will enable any discovery tools to dashboard a list of applications deployed, their version, and dependencies.

- **Application description customization**: Not all environments will hold the same configuration. For example, the replication count in a staging environment could be one, while in production, we may set up horizontal Pod scaling. The capability is also required when we want to inject dependencies enabling segregation of concerns.

Let's explore the tools available for application configuration management, keeping the capabilities in mind. The following section will deep dive into Helm, our first tool to explore.

Using Helm for application deployment

Helm is one of the popular configuration management tools in the Kubernetes ecosystem. It came into existence as early as 2015. It has come a long way in evolving itself and solving all the bottlenecks. Being a **Cloud Native Computing Foundation** (**CNCF**)-graduated project shows its maturity, production readiness, and value. Here are three key concepts of Helm:

- **Charts**: Charts are the basic units of applications in Helm. A chart is nothing but the bundled package of an application with all its dependencies.

- **Repository**: A bundled chart requires a consistent way of storage to distribute reliably, and repositories support this requirement. While open source applications can use a public repository, private repositories can be used for proprietary applications. Starting from Helm v3.8.0, any **Open Container Initiative (OCI)**-compliant repository will support Helm. This means that most container registries support Helm packages as well.

- **Release**: This is an instance of the chart running in the cluster. When we install a chart for the first time, it creates a new release version. Any update will be an increment in the release version. The construct enables release management capabilities such as rollout, rollback, and blue-green deployment.

Using Helm requires a client-side **command-line interface (CLI)** setup. In a macOS operating system, use `brew` to install the CLI, while the `choco` installer can be used for Windows. The following code snippet shows how to install the CLI on either of these operating systems:

```
# Helm install macOS
brew install Helm
# Helm install Windows
choco install kubernetes-helm
```

For more installation options, visit `https://helm.sh/docs/intro/install/`. We will explore Helm in two parts—the first part will cover working with an existing chart, and new chart development will be covered in the second part.

Working with an existing chart

Working with an existing chart can be categorized into repository management, release management, and cluster discovery. Here are a few repository management commands:

```
# Add a repo (Add bitnami repo to your local repo list)
helm repo add bitnami https://charts.bitnami.com/bitnami
# For private repository, use additional authentication flags
# To view the list of possible authentication flags
helm repo add --help

# Update the charts in all the added repo's
helm repo update
# Update the charts in a specific repo.
helm repo update bitnami
```

```
# Search for charts with the given name within the added repo
helm search repo wordpress
# Search charts in ArtifactHub, a famous open-source repo
helm search hub wordpress

#List all the repositories added
helm repo list
```

The preceding repository management commands are good enough for our day-to-day repository operations. Next, we will look at a few release management Helm commands, as follows:

```
# Install a chart
# Format
helm install <release-name> <chart-name>
# Example
helm install redis bitnami/redis
# Each chart will support a list of variables to be set
# Variables can be hierarchical. For example, the 'enabled' flag
is under the 'auth' hierarchy in the bitnami/redis chart.
helm install redis bitnami/redis --set auth.enabled=false
# Install with a value set in the values file
helm install redis bitnami/redis -f values.yaml
# If the same variable is present in both the command line set
and value file, the command line set takes precedence.
helm install redis bitnami/redis -f values.yaml --set auth.
enabled=false

# Update a release
# Format
helm upgrade <release-name> <chart-name>
```

```
# Example
helm upgrade redis bitnami/redis -f values.yaml --set auth.
enabled=true
# Roll back a release
# Format
helm rollback <release-name> <release>
# Example, roll back redis to the first release version
helm rollback redis 1


# Uninstall format
helm uninstall <release-name>
# Uninstall redis release
helm uninstall redis
```

The preceding commands are some frequently used release management commands. The following snippet will cover a couple of cluster discovery commands:

```
# List all the Helm releases in the target cluster
helm list
# Find the status of a release named redis
helm status redis
```

It's time to explore new chart development through a step-by-step, hands-on example.

Hands-on chart development

Helm charts are nothing but a set of configuration templates with variable placeholders in the templates. These placeholders can be replaced with values when templates are rendered for installation. Helm has a powerful **domain-specific language (DSL)** providing a wide range of constructs for variable replacement. We will look at some frequently used constructs to learn chart development in the upcoming sections.

Chart generation

A Helm chart bundle has a set of organized files and folders. Either we need to understand the structure to develop it from scratch or we can use the generator. In the hands-on example, we will use the generator to create a chart named `hello-world` (`helm create <chart-name>`), as illustrated in the following screenshot:

```
arunramakani@Aruns-MacBook-Pro helm % helm create hello-world
Creating hello-world
arunramakani@Aruns-MacBook-Pro helm % tree
.
└── hello-world
    ├── Chart.yaml
    ├── charts
    ├── templates
    │   ├── NOTES.txt
    │   ├── _helpers.tpl
    │   ├── deployment.yaml
    │   ├── hpa.yaml
    │   ├── ingress.yaml
    │   ├── service.yaml
    │   ├── serviceaccount.yaml
    │   └── tests
    │       └── test-connection.yaml
    └── values.yaml

4 directories, 10 files
```

Figure 9.1 – Creating a chart

Let's look at the use of each file, as follows:

- `Chart.yaml`: This is a file that holds a description for the chart. It contains information such as supported Helm version, chart version, application version, application name, description, other dependent charts, maintainers, and so on. It also has an attribute called `type` that holds the value of either `application` or `library`. `application` refers to the fact that we are packing a Kubernetes application, and `library` means that the chart contains utility functions for reuse.

- `charts` folder: This is a folder that can hold a list of dependent sub-charts. We can use sub-charts for many reasons. Dividing an extensive application into small modules with a sub-chart for each module is one way to use it. Another way could be to use it as a packaging mechanism for application dependencies such as a database. We could also use it as a holder of reusable functions that can be used as a shared library. An important thing to note here is that sub-charts can be independently deployed, meaning they cannot explicitly refer to the parent, but a parent can override values of sub-chart templates when required.

- `values.yaml`: This **YAML Ain't Markup Language** (**YAML**) file holds values that need to be replaced when templates are rendered. We can override this file with a new file through the CLI—for example, `helm upgrade redis-install-1 bitnami/redis -f values.yaml`. Also, we can use the `set` flag in the CLI to override a specific value.

- `template` folder: `NOTES.txt`, `_helpers.tpl`, and Kubernetes resource YAML templates are files that are part of the `template` folder. The `NOTES.txt` file is a template file that will be rendered and printed in the console when we run `helm install` or `helm upgrade`. The `_helpers.tpl` file will hold reusable functions that can be used across a chart. The rest of the files will be standard Kubernetes resource templates relevant to the application. Using the Helm `create` command to generate a basic chart adds a list of Kubernetes resources required for application deployment into the `template` folder. We can delete resources that are not necessary.

- `tests` folder: This can hold unit tests to test the logic we write in the resource template.

Helm uses the `template` package from the Go language and provides many powerful templating constructs to render complex scenarios. The following sections will explain each concept using the `hello-world` example.

Variable access

When templates are rendered, we can replace placeholder variables by simply specifying the variable hierarchy with the following syntax:

```
# Variable reference syntax - {{ variable-hierarchy }}
# Examples
# Refer deployment.yaml line no 9
{{ .Values.replicaCount }}
# Refer _helpers.tpl line no 50
{{ .Release.Name }}
```

In the preceding example, note that `.` acts as a separator representing the variable hierarchy. We start with `.` representing the root, then refer to one of the root objects. Note that we could have a variable with a multiple-depth hierarchy—for example, `{{ .Values.image.repository }}`. Here are some important built-in root objects available for us to use:

- `Release`: This object holds release-related information such as the release's name, release namespace, revision number, and so on.

- `Values`: An object formed with the values file/command-line set flags.

- `Chart`: Values defined in the `chart.yaml` file will be available under this object.

There are more objects available, such as `Files`, `Capabilities`, `Template`, and so on. Refer to `https://helm.sh/docs/chart_template_guide/builtin_objects/` for a complete list.

> **Tip**
>
> To remove an object or a specific attribute from the template, use the `--set` command-line with a `null` value—for example, `--set livenessProbe.httpGet=null`.

Functions and pipelines

We may have requirements to replace variables after doing some transformation, and built-in functions and pipelines can help with these. For example, refer to the following code snippet:

```
# Refer _helpers.tpl line no 40
app.kubernetes.io/version: {{ .Chart.AppVersion | quote }}
```

We refer to the application version and then enclose a string inside a quote in the preceding example. There are two essential things to note here. The quote is an built-in function available for us to use, and `|` will help pipe the output from one instruction to another. Here is a list of some frequently used functions:

- `indent`: Useful to format the configuration YAML. It takes a numerical input and indents the row with the specified index.

- `nindent`: The function works like the `intent` function, with an addition of a newline at the beginning.

- `trunc`: Truncates a string with the specified number of indexes.

- `trimSuffix`: This method takes a string suffix as input and truncates the suffix if it is present in the operating string.

- `replace`: The `replace` method can replace one substring with another in an operating string.

- `semverCompare`: This function can be used to compare two semantic versions.

These are some of the functions used in `deployment.yaml` and `_helpers.tpl`. Refer to `https://helm.sh/docs/chart_template_guide/function_list/` to look at an extensive list of built-in functions.

> **Information**
>
> Note that there is - in many template placeholders—for example, the sixth line of `deployment.yaml` has `{{-`. This instructs the template engine to remove whitespace on the left. Similarly, `-}}` can remove whitespace on the right.

Flow control

Flow controls make any programming language powerful and enable us to encode complex logic. Helm's template language provides three flow controls. The first flow control available is the standard `if/else` statement. It is helpful to include a block based on a specific condition. The following code snippet from the `ingress.yaml` file checks the Kubernetes version to decide on the ingress **application programming interface (API)** version:

```
# Refer ingress.yaml line no 9
{{- if semverCompare ">=1.19-0" .Capabilities.KubeVersion.
GitVersion -}}
apiVersion: networking.k8s.io/v1
{{- else if semverCompare ">=1.14-0" .Capabilities.KubeVersion.
GitVersion -}}
apiVersion: networking.k8s.io/v1beta1
{{- else -}}
apiVersion: extensions/v1beta1
{{- end }}
```

Note that `Capabilities` is an built-in object providing the capabilities of the target Kubernetes cluster. The second flow control, `with`, allows us to create a block with a specific variable scope. Refer to the following code snippet from `serviceaccount.yaml`:

```
# Refer serviceaccount.yaml line no 8
# We create a scope block with variable Values.serviceAccount.
annotations
{{- with .Values.serviceAccount.annotations }}
annotations:
  {{- toYaml . | nindent 4 }}
{{- end }}
```

Remember our earlier discussion about variable references? We mentioned that the initial `.` refers to all objects' roots. Inside a `with` block, the definition changes. The initial `.` within the block will refer to the scope variable defined. The third flow control is the range used for looping. Refer to the following code snippet from `NOTES.txt`:

```
# Refer NOTES.txt line no 2
# Loops over the hosts with an inner loop on path
{{- range $host := .Values.ingress.hosts }}
  {{- range .paths }}
  http{{ if $.Values.ingress.tls }}s{{ end }}://{{ $host.host }}{{
.path }}
  {{- end }}
{{- end }}
```

We declare a new variable, `host`, in the preceding example, and refer to it within the loop. Similarly, we could use variable declaration in other places as well.

> **Tip**
> We can use `dry-run` and a `disable-openapi-validation` flag with `helm install` or `helm upgrade` to debug or validate the YAML outputs.

Named templates

Named templates are frequently used, and they act as static custom-defined functions. We define a template with a name and then import them into the required place. Generally, these named templates are described in the helper file and reused across the chart. Refer to the two pieces of code in the following snippet:

```
# Template named hello-world.selectorLabels is defined in the _
helpers.tps line no 45
# hello-world is a release name prefix added to avoid name
conflicts when we have sub-charts and dependent charts.
{{- define "hello-world.selectorLabels" -}}
app.kubernetes.io/name: {{ include "hello-world.name" . }}
app.kubernetes.io/instance: {{ .Release.Name }}
{{- end }}


# deployment.yaml line number 12
# Template included with the name
matchLabels:
   {{- include "hello-world.selectorLabels" . | nindent 6 }}
```

Note that the template output can be piped with other built-in functions. We have covered most of the skills required to create new Helm charts for our bespoke applications. After changing the image name to `hello-world` in the `value.yaml` file, we can deploy the chart. Refer to the following screenshot for the chart installation:

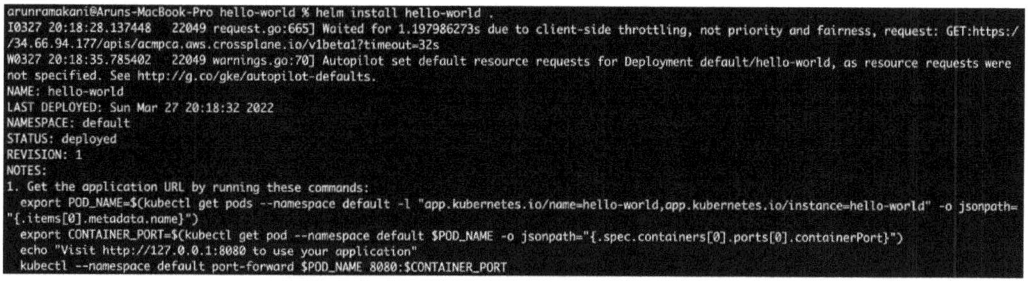

```
arunramakani@Aruns-MacBook-Pro hello-world % helm install hello-world .
I0327 20:18:28.137448   22049 request.go:665] Waited for 1.197986273s due to client-side throttling, not priority and fairness, request: GET:https:/
/34.66.94.177/apis/acmpca.aws.crossplane.io/v1beta1?timeout=32s
W0327 20:18:35.785402   22049 warnings.go:70] Autopilot set default resource requests for Deployment default/hello-world, as resource requests were
not specified. See http://g.co/gke/autopilot-defaults.
NAME: hello-world
LAST DEPLOYED: Sun Mar 27 20:18:32 2022
NAMESPACE: default
STATUS: deployed
REVISION: 1
NOTES:
1. Get the application URL by running these commands:
  export POD_NAME=$(kubectl get pods --namespace default -l "app.kubernetes.io/name=hello-world,app.kubernetes.io/instance=hello-world" -o jsonpath=
"{.items[0].metadata.name}")
  export CONTAINER_PORT=$(kubectl get pod --namespace default $POD_NAME -o jsonpath="{.spec.containers[0].ports[0].containerPort}")
  echo "Visit http://127.0.0.1:8080 to use your application"
  kubectl --namespace default port-forward $POD_NAME 8080:$CONTAINER_PORT
```

Figure 9.2 – Installing the chart

In the next section, we can use Kustomize to customize the configuration.

Customizing configurations with Kustomize

Be it a configuration managed by Helm or other configuration management tools, Kustomize is one of the best tools for configuration customization. Let's look at some of the use cases for Kustomize, as follows:

- Keeping environment-specific customization separate from the base configuration is one use case. For example, replication counts can be done in staging, while the production environment could be enabled with auto-scaling.

- Managing cross-cutting configurations outside the base configuration is another use case. For example, the application operator working with governance-specific labels in all deployments can keep the configuration separate from the base configuration. It can enable **separation of concerns** (**SoC**) for multi-persona collaboration without friction. Injecting a service mesh configuration as a cross-cutting concern is another example.

- The third use case is fixing vulnerabilities as a step in the configuration pipeline. Consider that there is a security vulnerability with an nginx image. The security team can add a customization step in the pipeline to ensure that the vulnerable version of nginx is updated for all deployments.

- The classic use case is to avoid abstraction leaking, as we discussed many times in the previous chapters. When we want to reuse the base configuration template across a few similar workloads, we can consider Kustomize as the new parameter.

The following screenshot represents how Kustomize patching can be used in a multi-persona collaboration environment:

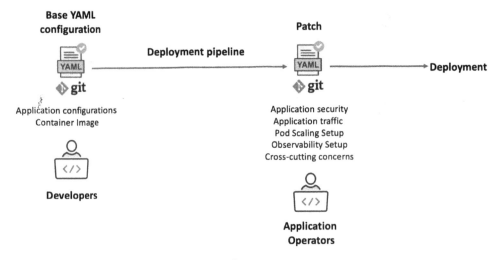

Figure 9.3 – Configuration customization

Let's look at a simple example to use Kustomize. We should have the base configuration on one side and kustomization.yaml on the other. kustomization.yaml defines how to customize the base configuration. Here is a sample kustomization.yaml file:

```
resources:
- deployment.yaml
commonLabels:
     team-name: alpha
namespace: test
```

The preceding configuration refers to the base configuration and defines how to customize it. The deployment.yaml file is the base configuration reference under the resources section. The commonLabels configuration adds the team's name as a label to the deployment, and the namespace configuration will override the deployment resource namespace. We can run the kubectl kustomize . command to perform the customization. Refer to the following screenshot:

```
arunramakani@Aruns-MacBook-Pro patchesStrategicMerge % kubectl kustomize .
apiVersion: apps/v1
kind: Deployment
metadata:
  labels:
    app: busybox
    team-name: alpha
  name: busybox-deployment
  namespace: test
spec:
```

Figure 9.4 – Configuration customization (continued)

Kustomize can work with Crossplane configurations as well. An example to add a label to the composition is available in the hands-on example repository for this chapter.

> **Tip**
>
> helm install can use Kustomize as a post-render step by specifying the path to kustomization.yaml. The syntax is helm install <release-name> <chart-name> --post-renderer ./path/to/executable.
>
> An example of using Helm and Kustomize is available at https://github.com/thomastaylor312/advanced-helm-demos/tree/master/post-render.

More than labels and namespaces, a lot more is possible with Kustomize. Refer to https://kubectl.docs.kubernetes.io/references/kustomize/kustomization/ for a deep dive into all possible customizations. This takes us to the end of the discussion on Kustomize, and in the following section, we will discuss KubeVela for application workload deployment.

Deploying application workloads with KubeVela

As discussed earlier, KubeVela is a project like Crossplane but focuses primarily on bespoke application workload. It can also cover off-the-shelf components via add-ons. Before getting into the details, let's look at ways to install KubeVela. We will do the KubeVela installation in two steps. The first part is installing the KubeVela CLI. We can use Homebrew or a script if you have a macOS operating system. In the case of Windows, we can use PowerShell. Here are the CLI installation instructions:

```
# Installing in macOS with Homebrew
brew update
brew install kubevela


# Installing in macOS with script
curl -fsSl https://kubevela.io/script/install.sh | bash -s 1.3.0


# Installing in windows with a script
powershell -Command "iwr -useb https://kubevela.io/script/install.
ps1 | iex"
```

As the next step, we should install KubeVela into the Kubernetes cluster, which is nothing but a set of **Custom Resource Definitions (CRDs)**. Here are the KubeVela CRDs' installation instructions. We can use either the CLI or a Helm chart:

```
# Using the CLI
vela install


# Using a Helm chart
helm repo add kubevela https://charts.kubevela.net/core
helm repo update
helm install --create-namespace -n vela-system kubevela
kubevela/vela-core --version 1.3.0 --wait
```

Additionally, we can enable add-ons. Add-ons enhance the capability of KubeVela. For example, we can use the `velaux` add-on as an application management dashboard. `terraform-gcp` is another add-on useful to compose **Google Cloud Platform** (**GCP**) resources' dependencies:

```
# View list of Add-ons available
vela add-on list
# Install the application management dashboard add-on
vela add-on enable velaux
```

We are all good to start using KubeVela. The core of the KubeVela configuration is the application API, and the anatomy of the application API is described in the following section.

Anatomy of a KubeVela application definition

The application specification carries the following four key sections:

- **Components**: Components are the root specification representing what we want to deploy as a Kubernetes workload. It can also be an external off-the-shelf dependency. In the case of Kubernetes workloads, it can be a Helm chart or a Kustomize package, or other Kubernetes objects such as `Deployment` and `Job`. The external off-the-shelf dependency components can be a Terraform module, CloudFormation template, or even a Crossplane **Composite Resource** (**XR**)/Claim.

- **Traits**: Traits are nothing but declarative operational behavior. Application rollout behavior, auto-scaling rules, and route rules are some examples of a trait. Traits are attached to the components, and we could have more than one trait attached.

- **Policies**: Policies are a set of rules to be enforced. Pod security policy and health-check configurations are a couple of examples.

- **Workflow**: A workflow is a final section that allows us to control the component delivery process. Approval steps and environment-specific traits are examples.

Have a look at the complete list of components, traits, policies, and workflows supported by our cluster using the following commands:

```
# List of Components
kubectl get ComponentDefinition -n vela-system
# List of Traits
kubectl get TraitDefinition -n vela-system
# List of Policies
```

```
kubectl get PolicyDefinition -n vela-system
# List of Workflows
kubectl get WorkflowStepDefinition -n vela-system
```

The following screenshot describes the anatomy of a KubeVela application with a hands-on example:

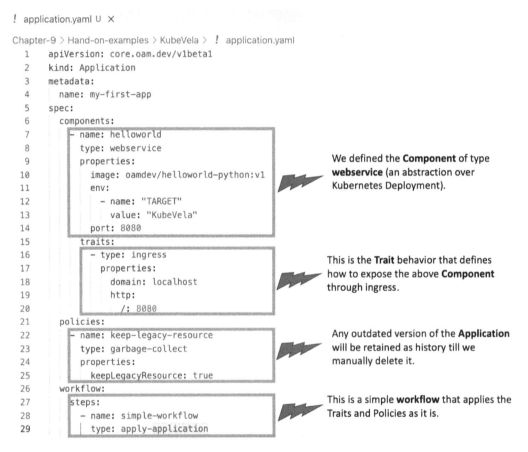

Figure 9.5 – Application API

Apply the application YAML, and you will find that our hello-world application is running successfully, as we can see here:

Figure 9.6 – Deployed application

The KubeVela community has developed many components, traits, policies, and workflows based on the **Open Application Model (OAM)** specifications and documented them in the usage examples. The list will cover most of the requirements. Refer to `https://kubevela.io/docs/end-user/components/references` for a deep dive. If we have a custom requirement, KubeVela has all the ingredients to compose a custom component, trait, policy, and workflow. It is again creating and registering new CRDs. As with Crossplane, KubeVela also provides a framework to develop these CRDs. That takes us to the end of the chapter and the discussion on KubeVela.

Summary

The chapter discussed some popular configuration management tools from the perspective of bespoke applications. While we did not cover every aspect of the tool, we did cover basic concepts, usage patterns, and a hands-on example. Each tool discussed here requires a book by itself to learn about in depth, which is beyond the scope of this book. The concepts, patterns, and tools we discussed will guide us to approach **end-to-end (E2E)** automation of the whole application using Crossplane.

The next chapter will go through a hands-on journey to onboard a complete application and its dependency using Crossplane, Helm, and a few other tools.

10

Onboarding Applications with Crossplane

This will be a fully hands-on chapter where we will look at the end-to-end automation of an application and all its dependencies. The dependencies will involve the setup of the project repository, creating the **Continuous Integration and Continuous Deployment (CI/CD)** pipelines, dependent infrastructure resources, and so on. You will see the real power of how Crossplane can automate every possible step, starting from the initial repository setup. We will go through the hands-on journey from the perspective of three different personas. The three personas are the platform developer creating the required XR/claim APIs, the application operator configuring the application deployment using the XR/claim, and the developer contributing to the application development. The platform developer persona is the key to the whole journey, so most of the content in this chapter will be from their perspective. Whenever required, we will explicitly mention the other personas. The hands-on journey will cover application, services, and infrastructure, all three aspects of automation with Crossplane.

The following are the topics covered in this chapter:

- The automation requirements
- The solution
- Preparing the control plane
- Automating the application deployment environment
- The repository and CI setup
- The deployment dependencies
- API boundary analysis

We will start with the requirement from the product team to explore the ways to automate.

The automation requirements

We will start our high-level requirement story from the perspective of an imaginary organization, *X*. They are planning to develop a new e-commerce website named `product-a`. It has many modules, each functional at a different time in the customer journey, for example, cart, payment, and customer support. Each model requires independent release and scaling capabilities while sharing a standard website theme and a unified experience. The product architecture group has recommended micro-frontend architecture with separate deployment for each module in Kubernetes. They also suggested that an individual team will develop the website framework, shared UI components, and cross-cutting concerns in the form of a library. The independent module team can use these dependent libraries to build their features. The product team has recently heard about Crossplane and its ability to automate the applications from end to end. They wanted to use the opportunity of developing a greenfield product and experiment with Crossplane to set up a high-velocity, reliable product development practice. They have reached the platform team, requesting help to develop a **proof of concept** (**POC**). The POC project will be the scope of our hands-on journey in this chapter. The following diagram represents what the product development team wanted to achieve:

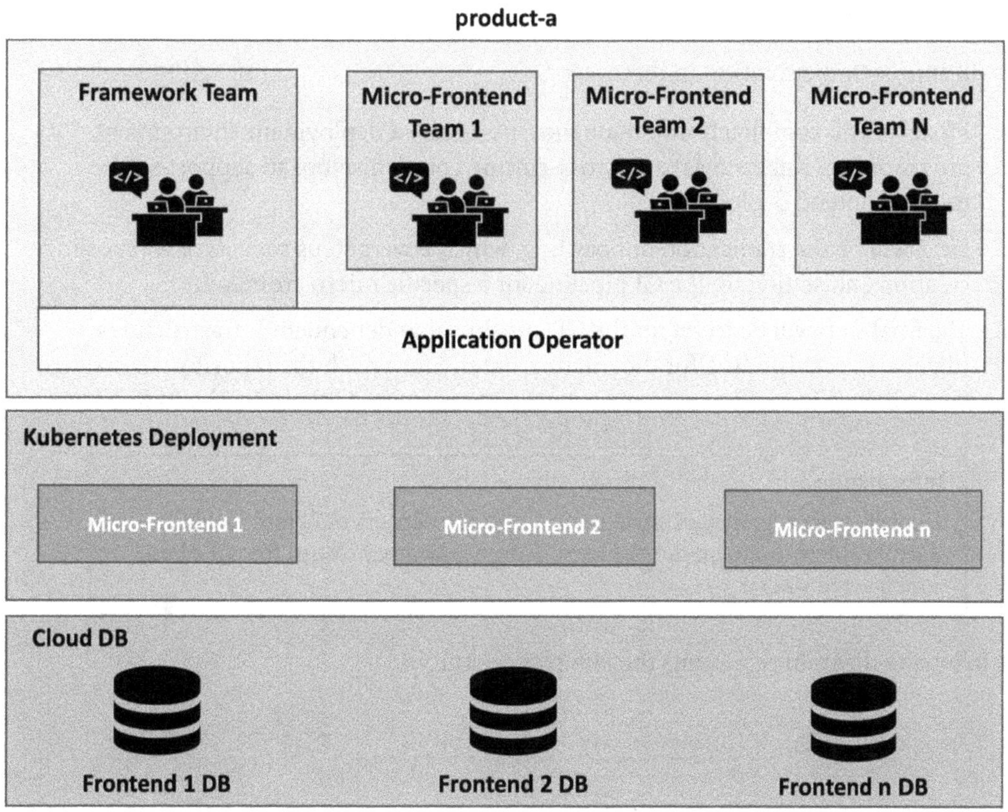

Figure 10.1 – Product team requirements

> **Information**
> Please note that both the requirements and solutions discussed in the chapter are not exhaustive. Our attempt here is to look for ways to approach automation from end to end, covering the entire application life cycle and its dependencies.

The following section explores one possible solution option from the perspective of a platform engineer using Crossplane.

The solution

We will approach the solution in three steps:

1. First, we will completely automate the `product-a` deployment environment provisioning (Kubernetes) and cross-cutting concern setups to support all the micro-frontend deployment.

2. Next will be the application onboarding, which covers steps such as new repository creation and setting up the CI pipeline for a specific micro-frontend.

3. The final step will be to set up the CD pipeline and dependent infrastructures (database) provisioning for the micro-frontend for which the repository is created. We will do this using a set of providers, such as Helm, GitLab, GCP, and Kubernetes.

> **Information**
>
> We will create a template GitLab project with the dependent library and kick-start the micro-frontend development using a repository cloned from the base template repository.

The following diagram represents the complete solution:

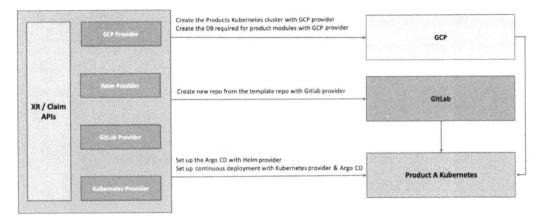

Figure 10.2 – High-level solution view

The following stages cover the high-level solution in the preceding diagram in a bit more detail:

- **Stage 1**: The first stage will be to get the Crossplane control plane ready with the required providers (GCP, Helm, Kubernetes, and GitLab) and configurations.

- **Stage 2**: The next stage will create a Kubernetes cluster to deploy the micro-frontend using the GCP provider. We will also immediately create the Helm and Kubernetes provider configuration in the control plane cluster, referring to the `product-a` cluster. The Helm provider configuration helps to set up Argo CD in the `product-a` cluster. The Kubernetes provider configuration will help deploy micro-frontend applications into the `product-a` cluster.

- **Stage 3**: The third stage is relevant to every micro-frontend application in the product. This step will create a new repository for the micro-frontend from the template repository. While creating the new repository, we will also clone the CI pipeline.

- **Stage 4**: The final stage will be to set the CD for the created repository using the Kubernetes provider. The Kubernetes provider configuration created in stage 2 will be used here. The stage will also create the cloud database instance required by the submodule/micro-frontend.

The rest of the chapter will investigate details of how we configure Crossplane and implement the solution discussed. The following section will deep dive into the control plan setup required to implement the use case.

> **Information**
>
> The complete example is available at `https://github.com/PacktPublishing/End-to-End-Automation-with-Kubernetes-and-Crossplane/tree/main/Chapter10/Hands-on-example`.

Preparing the control plane

This is the stage to install the required components into the Crossplane cluster. We will establish the necessary providers and respective configurations. The first step will be to install the GCP provider.

The GCP provider

This is the same step we took in *Chapter 3*, *Automating Infrastructure with Crossplane*, but slightly deviating from it. We will differ in how we create and use the GCP provider configuration. It is good to have an individual provider configuration for each product team to enhance security, auditing, policy compliance, governance, and so on in using the XR/claim APIs. Each product team and platform team should create a different provider configuration referring to a separate GCP service account secret. The provider configurations will be named against the product (product-a), and a new namespace will be created with the same name. The compositions will be developed in such a way to refer to the provider configuration based on the claim namespace dynamically. It is one of the multi-tenancy patterns we discussed in *Chapter 7*, *Extending and Scaling Crossplane*. To finish the GCP setup, do the following:

1. Execute GCP-Provider.yaml to install the provider. Wait until the provider pods are up and running.

2. Meanwhile, ensure that the Kubernetes Secret with the product-a GCP service account is available in the cluster. This Secret will be referred to in the provider configuration. To remind yourself of the steps to make the Secret, refer to the *Configure the provider* section in *Chapter 3*.

3. Once you have the Secret available, execute Provider-Config.yaml to create the product-specific provider configuration. Note that the name of the provider configuration is product-a.

4. Finally, apply namespace.yaml to create the product-a namespace. It is an additional step to hold Claim objects.

The preceding steps will ensure that the GCP provider is fully set. In the following section, we will look at the GitLab provider.

The GitLab provider

We will use the GitLab provider to manage the micro-frontend repository and CI pipeline. The free account provided by GitLab is good enough to continue with our experiment. The provider setup is done in three steps:

1. **GitLab credentials**: We need to create the GitLab access token as a Kubernetes Secret. It will be referred to in the GitLab provider configuration. Generate a GitLab access token in the GitLab UI user setting. Use the following command to create the Secret:

    ```
    # Create Kubernetes secret with the access token
    kubectl create secret generic gitlab-credentials
    ```

```
-n crossplane-system --from-literal=gitlab-
credentials=<YOUR_ACCESS_TOKEN>
```

2. **Installing the provider**: Execute `provider-gitlab.yaml` to install the GitLab provider and wait until the pods are up and running.

3. **Configuring the provider configuration**: Execute `provider-config.yaml` to create the provider configuration. Again, it will be specific to the product with the name `product-a`.

We are done with the GitLab provider setup. The following section will look at the Helm and Kubernetes provider setup.

Helm and Kubernetes provider setup

Both the Helm and Kubernetes providers are helpful to configure a remote or the same Kubernetes cluster. It is the remote Kubernetes cluster created for `product-a` in our case. Both providers require credentials to access the remote cluster. The product-specific provider configuration will be created automatically for the remote cluster when we provision the cluster with our XR API. We will look at more details on this in the next section. We will only install the provider for now. Execute `Helm-Provider.yaml` and `k8s-Provider.yaml` to install the providers. Refer to the following screenshot showing the installation of all providers and respective configuration setup:

```
arunramakani@Aruns-MacBook-Pro Hands-on-example % kubectl apply -f Step-1-ProviderSetup/Platform-OPS/GCP
provider.pkg.crossplane.io/provider-gcp created
arunramakani@Aruns-MacBook-Pro Hands-on-example % kubectl apply -f Step-1-ProviderSetup/Platform-OPS/GCP/product-a
providerconfig.gcp.crossplane.io/product-a created
namespace/product-a created
arunramakani@Aruns-MacBook-Pro Hands-on-example % kubectl apply -f Step-1-ProviderSetup/Platform-OPS/Helm
provider.pkg.crossplane.io/provider-helm created
arunramakani@Aruns-MacBook-Pro Hands-on-example % kubectl apply -f Step-1-ProviderSetup/Platform-OPS/Gitlab
provider.pkg.crossplane.io/provider-gitlab created
arunramakani@Aruns-MacBook-Pro Hands-on-example % kubectl apply -f Step-1-ProviderSetup/Platform-OPS/Gitlab/product-a
providerconfig.gitlab.crossplane.io/product-a created
arunramakani@Aruns-MacBook-Pro Hands-on-example % kubectl apply -f Step-1-ProviderSetup/Platform-OPS/k8s
provider.pkg.crossplane.io/provider-kubernetes created
```

Figure 10.3 – Provider setup

To run the setup yourself, use the following commands:

```
# GCP Provider
kubectl apply -f Step-1-ProviderSetup/Platform-OPS/GCP
kubectl apply -f Step-1-ProviderSetup/Platform-OPS/GCP/
product-a


# Helm Provider
kubectl apply -f Step-1-ProviderSetup/Platform-OPS/Helm
```

```
# GitLab Provider
kubectl apply -f Step-1-ProviderSetup/Platform-OPS/Gitlab
kubectl apply -f Step-1-ProviderSetup/Platform-OPS/Gitlab/
product-a


# Kubernetes Provider
kubectl apply -f Step-1-ProviderSetup/Platform-OPS/k8s
```

This takes us to the end of configuring the Crossplane control plane. All these activities are meant to be done by the platform team. In the following section, we will deep dive into setting up a remote Kubernetes cluster as a deployment environment for product-a.

Automating the application deployment environment

The complete Kubernetes cluster creation and configuring of the cross-cutting concerns will be automated using this step. We will develop an XR/claim API, which does the following:

1. Provisions a remote GKE cluster

2. Sets up Helm and the Kubernetes provider configuration for the GKE cluster

3. Installs Argo CD using the Helm provider into the product-a GKE cluster

Let's look at the XRD and composition to understand the API in detail (refer to the XRD and composition in the book's GitHub repository). We will capture two mandatory parameters (node count and machine size). The size parameter takes either BIG or SMALL as an enum value. Inside the composition, we have composed five resources. The following is the list of resources and their purpose:

- **Cluster and NodePool**: Cluster and NodePool are two relevant resources responsible for GKE cluster provisioning. It is very similar to the way we provisioned GKE in *Chapter 5, Extending Providers*. The node count and the machine type will be patched into the node pool. The node pool is again referred to inside the cluster. Both resources will refer to the GCP provider configuration dynamically using the claim namespace. Also, the Secret required to connect to the GKE cluster is stored in the claim namespace. Refer to the following code snippet on the patching operation in the cluster resource:

    ```
    patches:
    - fromFieldPath: spec.claimRef.namespace
    ```

```
    toFieldPath: spec.providerConfigRef.name
  - fromFieldPath: spec.claimRef.name
    toFieldPath: metadata.name
  - fromFieldPath: spec.claimRef.namespace
    toFieldPath: spec.writeConnectionSecretToRef.namespace
  - fromFieldPath: spec.claimRef.name
    toFieldPath: spec.writeConnectionSecretToRef.name
    transforms:
      - type: string
        string:
          fmt: "%s-secret"
```

- **Helm and Kubernetes ProviderConfig**: As the cluster is ready, it's time to create the Helm and Kubernetes provider configuration. The provider configuration will refer to the newly created cluster Secret. Another critical point is defining the readiness check as none, as ProviderConfig is not an external resource. Failing to do so will not allow the XR/claim to become ready. Refer to the following code snippet:

```
# Patches and reediness check from the Helm Provider
config
patches:
- fromFieldPath: spec.claimRef.namespace
  toFieldPath: spec.credentials.secretRef.namespace
- fromFieldPath: spec.claimRef.name
  toFieldPath: spec.credentials.secretRef.name
  transforms:
  - type: string
    string:
      fmt: "%s-secret"
- fromFieldPath: spec.claimRef.name
  toFieldPath: metadata.name
  transforms:
  - type: string
    string:
      fmt: "%s-helm-provider-config"
readinessChecks:
- type: None
```

- **Install Argo CD**: We will install Argo CD into the cluster using the Helm provider. Again, the provider configuration will be referred to dynamically with a predictable naming strategy. Argo CD is designed to enable CD for the micro-frontend repositories.

Information

Note that the cluster creation XR/claim API example discussed here is not production ready. You should be installing other cross-cutting concerns using the Helm or Kubernetes provider. Also, we missed many fine-grained cluster configurations. Refer to `https://github.com/upbound/platform-ref-gcp` for a more detailed cluster configuration.

To establish and validate our cluster API into the control plane, execute the following commands:

```
# Install GCP Cluster XR/Claim API
kubectl apply -f Step-2-CreateProductTeamsKubernetesCluster/
Platform-OPS
# Validate the health of installed API
kubectl get xrd
kubectl get composition
```

The platform team that manages the control plane will do the preceding operations. Refer to the following screenshot where the APIs are established:

```
arunramakani@Aruns-MacBook-Pro Hands-on-example % kubectl apply -f Step-2-CreateProductTeamsKubernetesCluster/Platform-OPS
composition.apiextensions.crossplane.io/xcluster-dev created
compositeresourcedefinition.apiextensions.crossplane.io/xgcpclusters.learn.unified.devops created
arunramakani@Aruns-MacBook-Pro Hands-on-example % kubectl get xrd
NAME                           ESTABLISHED   OFFERED   AGE
xgcpclusters.learn.unified.devops   True          True      23s
arunramakani@Aruns-MacBook-Pro Hands-on-example % kubectl get composition
NAME          AGE
xcluster-dev  39s
```

Figure 10.4 – Cluster API

As a next step, the application operator close to the product team can create the cluster using a claim configuration. The application operator will create a GKE cluster with the name `product-a` using the following commands:

```
# Create the GCP Cluster using a Claim object
kubectl apply -f Step-2-CreateProductTeamsKubernetesCluster/
Application-OPS
# Validate the health of the GKE cluster and the Argo CD
```

```
kubectl get GCPCluster -n product-a
kubectl get release
```

Refer to the following screenshot where the GKE cluster and Helm releases are established:

```
arunramakani@Aruns-MacBook-Pro Hands-on-example % kubectl apply -f Step-2-CreateProductTeamsKubernetesCluster/Application-OPS
gcpcluster.learn.unified.devops/product-a-cluster created
arunramakani@Aruns-MacBook-Pro Hands-on-example % kubectl get GCPCluster -n product-a
NAME                READY   CONNECTION-SECRET   AGE
product-a-cluster   True                        6m3s
arunramakani@Aruns-MacBook-Pro Hands-on-example % kubectl get release
NAME                           CHART       VERSION   SYNCED   READY   STATE      REVISION   DESCRIPTION        AGE
product-a-cluster-helm-argocd  argo-cd     2.3.3     True     True    deployed   1          Install complete   10m
```

Figure 10.5 – Cluster claim

We are all good with the cluster creation. We will discuss the next stage to onboard the micro-frontend repository in the following section.

The repository and CI setup

At this stage, an XR/claim is developed to clone the template repository to create the new micro-frontend repository and CI pipeline. We can do this in two steps. First, we will configure GitLab, and then we'll develop an XR/claim API.

GitLab configuration

We need to make the following one-time configurations in GitLab before we start the XR/ claim API development:

- **Create the template project**: We need to have a template repository from which we will make a new micro-frontend repository. You can access the template repository I have created at `https://gitlab.com/unified.devops/react-template`. The repository has a GitLab pipeline set up to build and push the Docker image into the Docker Hub registry. You can also set up a private registry here. We will automatically get the template project structure and CI set up while we clone the template repository for a micro-frontend. The Docker image name will be chosen based on the micro-frontend repository name.

- **Group for product-a**: We will keep all micro-frontend repositories in a single GitLab group to keep it organized, manage user permissions, and maintain environment variables for a CI pipeline. You can access the group I have created at `https://gitlab.com/unified-devops-project-x`.

- **Setup environment variables**: To enable the GitLab pipeline to access Docker Hub, we need to set up a couple of environment variables. We will add these variables at the group level so that all micro-frontend repository pipelines can access them. Go to the group-level settings CI/CD section. In the **Variables** section, add REG_USER and REG_PASSWORD with your Docker Hub credentials, as shown in the following screenshot:

Figure 10.6 – CI Docker Hub variables

> **Tip**
> Note that the group creation and user onboarding into the group can be automated. Considering doing that with Crossplane. An example of this is available at https://github.com/crossplane-contrib/provider-gitlab/tree/master/examples/groups.

We have all the components to develop our project onboarding XR/claim API. The following section will look at the details of the onboarding API.

The onboarding XR/claim API

If we look at the XRD (`gitproject-xrd.yaml`), we take in two parameters as inputs. The template's name refers to the template repository from which we should be cloning, and the group ID will determine the GitLab group under which the repository will be created. You can get the group ID from the GitLab group details page or group settings page. These two parameters make the API generic, so it can be used across the organization. The newly created micro-frontend repo URL and an access token to work with the repository will be stored as connection Secrets. We can use these with Argo CD to read the repo. Our example doesn't require the access token as the repository is public. It will be a simple composition to map the template name with a template URL, clone the repository into the specified group, and copy back the repository details into the Secret. The repository's name will be referred to from the name of the claim object. To establish and validate the onboarding API into the control plane, execute the following commands:

```
# Install the onboarding API
kubectl apply -f Step-3-GitProjectOnboarding/Platform-OPS
# Validate the health of installed API
kubectl get xrd
kubectl get composition
```

Refer to the following screenshot, where the APIs are established:

```
arunramakani@Aruns-MacBook-Pro Hands-on-example % kubectl apply -f Step-3-GitProjectOnboarding/Platform-OPS
compositeresourcedefinition.apiextensions.crossplane.io/xgitprojects.learn.unified.devops created
composition.apiextensions.crossplane.io/gitprojects-composition created
arunramakani@Aruns-MacBook-Pro Hands-on-example % kubectl get xrd
NAME                              ESTABLISHED   OFFERED   AGE
xgcpclusters.learn.unified.devops   True          True      18m
xgitprojects.learn.unified.devops   True          True      12s
arunramakani@Aruns-MacBook-Pro Hands-on-example % kubectl get composition
NAME                    AGE
gitprojects-composition  22s
xcluster-dev            18m
```

Figure 10.7 – Onboarding API

As a final step in the onboarding stage, the application operator can onboard the repository and CI pipeline using a `Claim` configuration. The application operator will create a repository with the name `micro-frontend-one` using the following commands:

```
# Create claim and validate
kubectl apply -f Step-3-GitProjectOnboarding/Application-OPS
kubectl get gitproject -n product-akubectl get xrd
```

Refer to the following screenshot where the claims are created in GitLab:

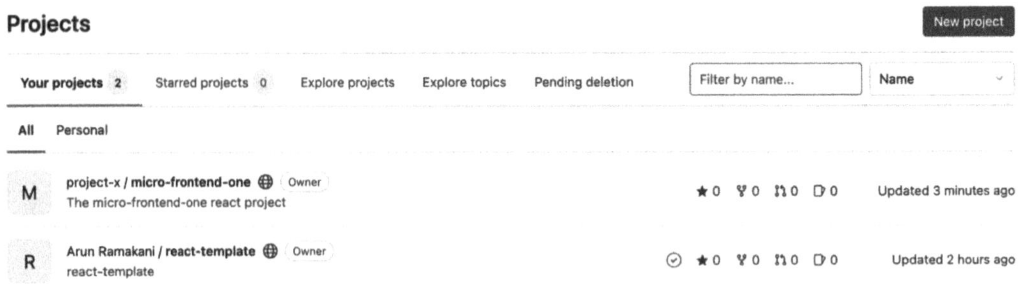

Figure 10.8 – Onboarding the repository

You can go to the CI/CD section of the new repository to run the CI pipeline to see that the Docker images are created and pushed into Docker Hub. Developers can now make changes to the repository, and any new commit will automatically trigger the GitLab CI pipeline. In the following section, we can investigate the final stage to set up CD and provision other infrastructure dependencies.

The deployment dependencies

The final stage is to automate the deployment dependencies for the micro-frontend. Automating the deployment dependencies means taking care of two aspects:

- **Infrastructure dependencies**: The step involves provisioning the needed infrastructure dependencies for the micro-frontend. In our case, we will create a GCP MySQL database. There could be more dependencies for an application. We will settle with just a database to keep the example simple.

- **Continuous deployment**: If you look at the `template-helm` folder inside our template repository (`https://gitlab.com/unified.devops/react-template/-/tree/main/template-helm`), it holds a Helm chart for deploying the application into Kubernetes. To deploy this Helm chart in a GitOps fashion, we must add an Argo CD configuration to the `product-a` Kubernetes cluster to sync the chart. We will construct an `Object`-type Kubernetes provider configuration, which can help apply any Kubernetes configuration to a target cluster. Our composition will compose an Argo CD configuration to deploy a Helm chart from a repository. Read more on how to use Argo CD for Helm deployment at `https://cloud.redhat.com/blog/continuous-delivery-with-helm-and-argo-cd`.

We will build a nested XR to satisfy the preceding requirement. The XWebApplication will be the parent API, and XGCPdb will be the nested inner XR. The parent API captures the product Git group and database size as input. The micro-frontend name will be another input derived from the name of the claim. The parent composition will compose the Argo CD config and an XGCPdb resource (inner XR). Refer to our example repo's application and database folder to go through the XRD and composition of both XRs. The following are a few code snippets that are key to understanding. In the Argo CD object, the following is the patch for the repository URL. We construct the GitLab URL from the group name and claim name (repository name). Look at the claim to see the actual input (Claim-Application.yaml). The following is the repository URL patch code:

```
- type: CombineFromComposite
  toFieldPath: spec.forProvider.manifest.spec.source.repoURL
  combine:
    variables:
    - fromFieldPath: spec.parameters.productGitGroup
    - fromFieldPath: spec.claimRef.name
    strategy: string
    string:
      fmt: "https://gitlab.com/%s/%s.git"
```

We dynamically patch the Kubernetes provider config name using a predictable naming strategy. The following is the code snippet for this:

```
- fromFieldPath: spec.claimRef.namespace
  toFieldPath: spec.providerConfigRef.name
  transforms:
    - type: string
      string:
        fmt: "%s-cluster-k8s-provider-config"
```

Another important patch is to bind the Docker image name dynamically. In our CI pipeline, we use the repository name as the Docker image name. As the claim name and the repository name are the same, we can use the claim name to dynamically construct the Docker image name. The following is the patch code snippet for this:

```
- fromFieldPath: spec.claimRef.name
  toFieldPath: spec.forProvider.manifest.spec.source.helm.
parameters[0].value
  transforms:
```

```
    - type: string
      string:
        fmt: "arunramakani/%s
```

source and destination are two key sections under the Argo CD config. This configuration provides information about the source of the Helm chart and how to deploy this in the destination Kubernetes cluster. The following is the code snippet for this:

```
source:
  # we just saw how this patched
  repoURL: # To be patched
  # The branch in which Argo CD looks for change
  # When the code is ready for release, move to this branch
  targetRevision: HEAD
  # Folder in the repository in which ArgoCD will look for
automatic sync
  path: template-helm
  helm:
    # We will patch our clime name here
    releaseName: # To be patched
    parameters:
    - name: "image.repository"
      # we just saw how this patched
      value: # To be patched
    - name: "image.tag"
      value: latest
    - name: "service.port"
      value: "3000"
destination:
  # Indicates that the target Kubernetes cluster is the same
local Kubernetes cluster in which ArgoCD is running.
  server: https://kubernetes.default.svc
  # Namespace in which the application is deployed
  namespace: # to be patched
```

To establish and validate our APIs in the control plane, execute the following commands:

```
kubectl apply -f Step-4-WebApplication/Platform-OPS/Application
kubectl apply -f Step-4-WebApplication/Platform-OPS/DB
kubectl get xrd
kubectl get composition
```

Refer to the following screenshot, where the APIs are established and validated:

```
arunramakani@Aruns-MacBook-Pro Hands-on-example % kubectl apply -f Step-4-WebApplication/Platform-OPS/Application
composition.apiextensions.crossplane.io/web-application-dev created
compositeresourcedefinition.apiextensions.crossplane.io/xwebapplications.learn.unified.devops created
arunramakani@Aruns-MacBook-Pro Hands-on-example % kubectl apply -f Step-4-WebApplication/Platform-OPS/DB
composition.apiextensions.crossplane.io/mysql created
compositeresourcedefinition.apiextensions.crossplane.io/xgcpdbs.learn.unified.devops created
arunramakani@Aruns-MacBook-Pro Hands-on-example % kubectl get xrd
NAME                                    ESTABLISHED   OFFERED   AGE
xgcpclusters.learn.unified.devops       True          True      35m
xgcpdbs.learn.unified.devops            True                    10s
xgitprojects.learn.unified.devops       True          True      16m
xwebapplications.learn.unified.devops   True          True      21s
arunramakani@Aruns-MacBook-Pro Hands-on-example % kubectl get composition
NAME                     AGE
gitprojects-composition  16m
mysql                    21s
web-application-dev      32s
xcluster-dev             35m
```

Figure 10.9 – Onboarding the application API

> **Tip**
> Note that we did not configure any access token for Argo CD to access GitLab
> as it is a public repository. We will have private repositories in real life, and
> a token is required. Refer to https://argo-cd.readthedocs.
> io/en/release-1.8/operator-manual/declarative-
> setup/#repositories to see how to set up an access token. Again, this
> can be automated as a part of repository onboarding.

As a final step in the application deployment automation stage, the application operator can provision the database as an infrastructure dependency and configure the CD setup using the following claim configuration:

```
apiVersion: learn.unified.devops/v1alpha1
kind: WebApplication
metadata:
  # Use the same name as the repository
  name: micro-frontend-one
  namespace: product-a
spec:
```

```
compositionRef:
  name: web-application-dev
parameters:
  # Group name in gitlab for the product-a
  productGitGroup: unified-devops-project-x
  databaseSize: SMALL
```

The application operator will use the following commands:

```
# Apply the claim
kubectl apply -f Step-4-WebApplication/Application-OPS
# Verify the application status, including the database and
ArgoCD config
kubectl get webapplications -n product-a
kubectl get XGCPdb
kubectl get object
```

Refer to the following screenshot, where the application infrastructure dependencies and CD configurations are provisioned:

Figure 10.10 – Onboarding the API

> **Tip**
> We have used Argo CD and Helm chart deployment to handle application automation. We can replace Helm with KubeVela, combine Helm/KubeVela with Kustomize, or even use a plain Kubernetes object as required for your team. Even Argo CD can be replaced with other GitOps tools, such as Flex.

This takes us to the end of the hands-on journey to automate the application from end to end. Our micro-frontend example and its dependent database are up and running now. In the following section of this chapter, we will discuss the reasoning behind our XR/claim API boundaries.

API boundary analysis

We divided the end-to-end automation into four stages. We can ignore stage one as it is about preparing the Crossplane control plane itself. It's essential to understand why we split the remaining stages into three with four XR/claim APIs. The following are the ideas behind our API boundaries:

- **The cluster XR/claim**: Setting up the cluster is not just relevant to `product-a`. All modern workloads are generally deployed in Kubernetes, and the organization will have many such cluster setup activities in the future. Building a separate API to enable reusability and centralized policy management makes sense. Another critical reason to keep the API separate is that the cluster setup is a one-time activity and acts as cross-cutting for further application workload deployments.

- **The onboarding API**: The XR/claim for the GitLab project onboarding is developed as a separate API. We don't need to onboard the repository and CI pipeline for every environment (production, staging, and development). That's why we decided to keep XGitProjectAPI/GitProject API separate.

- **The application API**: This is the step where we onboard the application infrastructure dependencies and CI setup, which is done once per environment. That's why we developed XWebApplication/WebApplication as a separate API. Note that there is an inner nested API for the database provisioning. The idea is to keep it separate as there are organization-wide policies in database provisioning. Note that the database API does not have a claim as it is designed to be used as only a nested API. The policy requirement is an assumption that may not be true for your case.

> Tip
> The repository URL and access token created with the onboarding API is required in the application API to set up CI. The onboarding API is a one-time activity, and the application API is used in every environment. If we have a different Crossplane for every environment (production, staging, and development), sharing the credentials across in an automated way could be challenging. Consider using an external key vault to sync the repository details from the onboarding API. Other Crossplane environments can synchronize these Secrets using tools such as External Secrets (`https://external-secrets.io/v0.5.3/`).

Summary

This chapter discussed one of the approaches to handling the end-to-end automation of applications, infrastructure, and services. There are multiple patterns to approach end-to-end control plane-based automation using the ways we learned throughout the book. I can't wait to see what unique ways you come up with. This chapter takes us to the end of learning Crossplane concepts and patterns and our hands-on journey.

In the final chapter, we will look at some inspirations to run a platform as a product. You will learn essential engineering practices that make our Crossplane platform team successful.

11
Driving the Platform Adoption

By now, you might be thinking about adopting Crossplane in your organization. As we've discussed many times, we must set up a platform team to support every product team across the organization adopting Crossplane. But many organizations fail with their technology platform projects because these platforms impact the agility of the consuming teams in many ways. Setting up proper boundaries and an interaction model between the product and platform teams is key to the success of a platform. This chapter aims to help you understand the aspects required to build and adopt a successful infrastructure platform. We will also look at a few common pitfalls that can lead to weak platform adoption and low return on investment.

The following are the topics covered in this chapter:

- Why we need an infrastructure platform as a product
- Understanding customers' needs
- The platform prodcut life cycle and team interaction
- The OAM personas

In the following section, we will start with understanding why we need to approach infrastructure platforms as product development.

Why we need an infrastructure platform as a product

We touched on this topic a bit in *Chapter 2*. It is time to look back and see why we need a platform team to develop the Crossplane APIs. The following are the three key reasons:

- **Cognitive load**: Any organization will tend to use a vast amount of cloud resources and other external services. These resources and services consist of tens of thousands of attributes to configure them according to the organization's requirements. Remembering the usage of each configuration attribute involves a significant cognitive load. Suppose we attempt to build this knowledge within the product team. In that case, the team will focus on technical complexity rather than product feature development, which is of direct business interest. If you look at the CNCF cloud-native landscape, it's vast (`https://landscape.cncf.io/`). Not every team can tame the terrain. It requires a specialized group to build this cognitive capability into the organization. Provided every product team faces the same situation, it makes sense to abstract this effort as a shared service managed by a platform team.

- **Policy management**: Infrastructure and other external services' usage policies come from security, compliance, product, and architecture teams. It requires a centralized team to track, consolidate, and encode the policies when these resources are provisioned and used. It won't be easy to keep the policies enforced and updated if we have these resources' automation as part of individual product teams.

- **Return on investment on reuse**: Generally, many tools, architecture patterns, and infrastructure setup patterns look similar across the organization. This is because of centralized IT strategies, the cross-pollination of successful designs, the organization's cognitive capability, and so on. These commonalities will translate into the reuse of infrastructure reference architecture. Provided we have the cognitive load and policy management advantages, the economics of return on investment with reuse is an added advantage.

Now that we are convinced that an infrastructure platform team is required, the following section will explore the product teams' expectations of the infrastructure platform team.

Understanding customers' needs

A platform team for infrastructure means external delivery dependency. This is something that is always avoided by agile product development teams. If you have worked in a large organization, you might have noticed that the product teams are often unhappy with what the platform team offers. They have issues with the delivery timeline, backlog coupling, and weak contracts. It is critical to set up a proper interaction model between the product and platform teams to mitigate these risks. The following are qualities that a platform team should possess to match the delivery expectations of product teams in the modern software engineering era:

- Product development practices
- Self-service
- Collaborative backlog management

The following section will look at the product development engineering practices for the platform team.

Product development practices

Adopting product development practices is a critical component that can make a platform team successful. The following are some of the product development practices for adoption:

- **Quick start**: Platform usage must be made quick and straightforward. Crossplane being an API-based platform solves part of the problem. APIs have a clear contract, and they are easy to understand. But this is not enough. We must make additional efforts to make the platform easy to adopt. We should have a quick-start guide, code repositories with examples, a support system to help adopting teams, and extensive API documentation. More importantly, the ability to quickly start should be a critical metric measured and continuously improved. We can do this with continuous feedback cycles between the product and platform teams. The feedback should be from multiple sources, such as customer surveys, usage statistics of various quick-starter asserts, usage statistics of the platform itself, and time to adoption. There should be incentives for the platform team to improve these metrics continuously.

- **Community of practice**: Building a community with participants from the platform and product teams is another essential aspect. Application operators could be natural participants of the community from the product team. Being close to the product team's requirements, they will bring a reality check to what we build as APIs in the platform. Again, there should be metrics and incentives to run the community successfully. We can use the community in many ways, such as for co-creation, as knowledge carriers, for two-way feedback flow channels, and for accelerating adoption. We will look at some of these options in detail in the following sections of the chapter.

- **Composable APIs**: One of the critical challenges with technology platforms is that they enforce a particular way of development that hinders innovation. The platform should have a basic set of fine-grained XRs, above which we can quickly compose the claim recipes required by the product team. It is recommended to have two-layered APIs. The first layer should have fine-grained XRs composing organization-centralized policies, and the following layer should be the recipe layer, composing product-level policies and requirements. We could have a basic set of reused recipes used across the organization and leave space for the community to innovate new recipes. Some organization setups may require Goldilocks governance to avoid the community's proliferation of too many new recipes.

> **Information**
>
> Goldilocks architecture governance has gained traction in modern software engineering practices where a balanced approach is taken between governance and flexibility.

In the following section, we will cover the self-service aspect of platform engineering.

Self-service

Self-service is one of the critical aspects of building a platform. Think about how any cloud provider works. We have the console and the APIs to create/operate the resource required. It's an entirely self-service model where users can manage resources according to their permissions. A refined resource granularity in a cloud provider environment will work for any organization and personas. We must have a similar experience for our platform as well. With Crossplane being an API-based resource composing platform, part of the problem is already solved. But we must put a lot of thought into defining the scope and granularity of the XR/claim APIs that the product teams will consume. What level of granularity will work for your organization is a crucial question to ask. It may depend on the organization's size, the team boundaries, the technical maturity of the developers/ platform team, and so on. Defining wrong boundaries will derail the self-service agenda.

Getting the boundary perfect at first is difficult. It keeps on improving in iterations by measuring the self-service metrics. In addition to API boundaries, we need to focus on the quick-start guide, technical support system, and building community, as discussed earlier, to enable self-service further. The whole experience should be like how a successful cloud provider platform works. The only difference is that we operate on a higher level of abstraction to tame the complexity.

Collaborative backlog management

A single centralized platform team serving multiple internal teams within the organization is not a fun place to work. We must manage the backlog to satisfy everyone's needs. There will be continuous requests from every direction to enhance an existing XR/claim or create a new one. We may end up in many unwanted situations, such as constant delivery pressure on the platform team, impact on the product team's delivery throughput, or frustrated product teams developing the capability locally. The following are some of the ways to mitigate these risks:

- **Backlog prioritization**: Score each backlog item with metrics such as time criticality, missed opportunity, and short-/long-term business value, in a prioritization session. Document the reason for each score for visibility. All product teams with backlog items should be actively involved in the prioritization session. There could be a tendency that each product team tries to push their score up in the prioritization sessions. We can tackle that with a scoring slider where a total score is set on a metric, and the sum of the scores for each backlog should not exceed the set value.

- **Expectation management**: Product teams are waiting for their backlog to be delivered, and they should be treated like end customers. It's essential to set expectations for the customers on when and what they can expect. Use Scrum of Scrums or any other scale agile framework to find a way to provide regular updates. Provide an update on the progress and the delivery date. Be open and transparent about any delayed delivery.

- **Community and governance**: Look at how open source projects such as Kubernetes and Crossplane work. They also face the same situation where there is a lot of expectation from many users for new features, enhancements, and bug fixes. These projects were initially started with a few focused developers. When the product moves from the alpha-beta stage to general availability, many customers start adopting it. At this point, managing the backlog becomes critical. These open source projects scale up the delivery velocity by building a solid community around them. Development activities by the community are governed by technical steering groups filled with core contributors. We could adopt the same model in the internal product as well. As we have said already, application operators close to the product team can be naturally attracted to the community. Product teams with pressing needs can allocate their application operator to co-create the product.

In the following section, we will investigate various stages of platform API development and how they impact the interaction between the platform team and the product team.

The platform product life cycle and team interaction

We may start our Crossplane API development as a small proof of concept working with one of the product teams. Later, when we are comfortable with the proof of concept, we will release an alpha version asking a few teams to consume and test it. Again, when we are satisfied with the alpha version, we will ask a few more sections of the organization to adopt and go live with the API. If everything goes well, we can move the API to general availability to be consumed by anyone in the organization. It will not end there. Sometimes, we will have the requirement to deprecate the APIs and mark them for removal in the future. The following figure represents the API journey:

Figure 11.1 – Crossplane API life cycle

Every API may not go through all the phases mentioned. It may take its own path, moving from left to right and skipping some stages depending on the situation.

> **Information**
>
> Sometimes, we may move directly to the deprecated state from alpha or beta if we find no business value or technical feasibility.

Each step requires a different engagement model and developer capability. The following list describes each of the life cycle phases:

- **Proof of concept and alpha stages**: Two things are essential to consider at these stages. First, the interaction model between the product and platform teams should be collaboration. This means both teams should work closely together as much as possible. The capability of the developers involved should be rapid innovators. We don't need the API to be scalable or reliable at this point. Our only aim should be to experiment with the feasibility and value as quickly as possible.

- **Beta stage**: This stage should continue the collaborative work format, but we should slowly move toward self-service. The product teams should be able to request access to the new API using some self-service portal and start using it. Creating self-service artifacts should be performed at this stage. The platform team should perfect the self-service portal and API in close collaboration with the product team. The capability of the developers involved should be a mix of rapid innovators and people who can bring stability/reliability. Our aim should be to get ready for organization usage.

- **General availability**: This is the stage of organization-wide usage. Two key aspects are being entirely self-service and having a stable and reliable API.

- **Depreciation and removal stage**: Sometimes, we need to deprecate and remove support for an API. Pushing the product team to get rid of the API usage is not easy because they are stuck with their priority. The platform team should use the facilitating interaction model to execute the activity. The platform developers perform the actual work for the product team to get rid of an API. Again, the platform developers at this stage are required to be rapid innovators.

> **Acknowledgment**
>
> Many of the concepts we discussed in this chapter are influenced by the ideas found in `https://teamtopologies.com/`, Evan Bottcher's work on platform engineering (`https://martinfowler.com/articles/talk-about-platforms.html`), and finally, some experience in implementing these concepts at work.

The following diagram represents the preceding API stages and the team interaction model discussed:

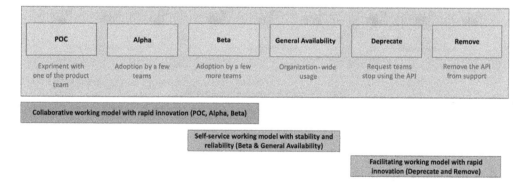

Figure 11.2 – API stages and team interaction model

This takes us to the end of the platform life cycle discussion. In the following section, we will look at the **Open Application Model (OAM)** specification to take some inspiration for the organization-level team structure from the perspective of platform engineering.

The OAM personas

We've touched upon OAM specifications several times throughout this book. We will refresh the same topic from the perspective of organizational structure. We will take some inspiration from the OAM model to organize the platform and its ecosystem. OAM proposes the following three personas to deploy and manage cloud-native applications:

- **Application developer**: Concentrates on application development, keeping the entire emphasis on developing features that add value to customers directly.

- **Application operator**: Offloads the complexity of configuring the applications in the cloud-native ecosystem from application developers. Enables the application development team to move faster with feature development. Application operators contribute to the end consumer indirectly.

- **Infrastructure operator**: Offloads the complexity of configuring the cloud, other infrastructure, and services across the organization. It allows the application operator to focus on configuring and operating the application.

Overlapping the OAM personas and other concepts we learned in this chapter, we could structure the platform team and its ecosystem as illustrated in the following diagram:

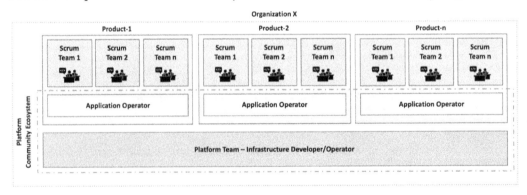

Figure 11.3 – Platform ecosystem

The following aspects explain the team constructs represented in the preceding diagram:

- The organization *x* is split into multiple products.

- Each product is divided into multiple Scrum teams handling different subdomains.

- One or more application operators support configuring and operating all applications from a given product.

- The platform team acts as a cross-cutting service to support all product teams for infrastructure, service, and application automation.

- The platform team and application operator form the community to support activities such as co-creation and accelerating adoption.

We could explore more models to structure the platform and its ecosystem. But the core idea is to manage the platform development as a product engineering practice treating all internal consumers as customers.

Summary

We started the book with an idea to support cloud-native architects, platform engineers, infrastructure operators, and application operators interested in simplifying infrastructure, application, and service automation with the Kubernetes APIs. Throughout the book, we've covered different aspects of building a fully declarative self-service control plane to automate infrastructure, applications, and services using Crossplane and Kubernetes. I hope you learned about the required concepts while having fun. Let's together take software engineering in to a new era.

Index

Packt.com

Subscribe to our online digital library for full access to over 7,000 books and videos, as well as industry leading tools to help you plan your personal development and advance your career. For more information, please visit our website.

Why subscribe?

- Spend less time learning and more time coding with practical eBooks and Videos from over 4,000 industry professionals

- Improve your learning with Skill Plans built especially for you

- Get a free eBook or video every month

- Fully searchable for easy access to vital information

- Copy and paste, print, and bookmark content

Did you know that Packt offers eBook versions of every book published, with PDF and ePub files available? You can upgrade to the eBook version at packt.com and as a print book customer, you are entitled to a discount on the eBook copy. Get in touch with us at customercare@packtpub.com for more details.

At www.packt.com, you can also read a collection of free technical articles, sign up for a range of free newsletters, and receive exclusive discounts and offers on Packt books and eBooks.

Other Books You May Enjoy

If you enjoyed this book, you may be interested in these other books by Packt:

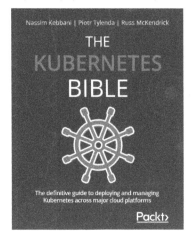

The Kubernetes Bible

Nassim Kebbani, Piotr Tylenda, Russ McKendrick

ISBN: 9781838827694

- Manage containerized applications with Kubernetes
- Understand Kubernetes architecture and the responsibilities of each component
- Set up Kubernetes on Amazon Elastic Kubernetes Service, Google Kubernetes Engine, and Microsoft Azure Kubernetes Service
- Deploy cloud applications such as Prometheus and Elasticsearch using Helm charts
- Discover advanced techniques for Pod scheduling and auto-scaling the cluster
- Understand possible approaches to traffic routing in Kubernetes

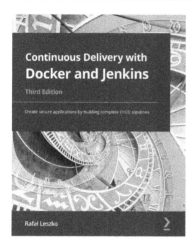

Continuous Delivery with Docker and Jenkins - Third Edition

Rafał Leszko

ISBN: 9781803237480

- Grasp Docker fundamentals and dockerize applications for the CD process
- Understand how to use Jenkins on-premises and in the cloud
- Scale a pool of Docker servers using Kubernetes
- Write acceptance tests using Cucumber
- Run tests in the Docker ecosystem using Jenkins
- Provision your servers and infrastructure using Ansible and Terraform
- Publish a built Docker image to a Docker registry
- Deploy cycles of Jenkins pipelines using community best practices

Packt is searching for authors like you

If you're interested in becoming an author for Packt, please visit authors. packtpub.com and apply today. We have worked with thousands of developers and tech professionals, just like you, to help them share their insight with the global tech community. You can make a general application, apply for a specific hot topic that we are recruiting an author for, or submit your own idea.

Share Your Thoughts

Now you've finished *End-to-End Automation with Kubernetes and Crossplane*, we'd love to hear your thoughts! Scan the QR code below to go straight to the Amazon review page for this book and share your feedback or leave a review on the site that you purchased it from.

https://packt.link/r/1-801-81154-7

Your review is important to us and the tech community and will help us make sure we're delivering excellent quality content.

www.ingramcontent.com/pod-product-compliance
Lightning Source LLC
Chambersburg PA
CBHW060541060326
40690CB00017B/3565